African farm
management

African farm
management

MARTIN UPTON

Department of Agricultural Economics and Management, University of Reading

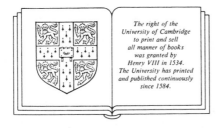

The right of the
University of Cambridge
to print and sell
all manner of books
was granted by
Henry VIII in 1534.
The University has printed
and published continuously
since 1584.

CAMBRIDGE UNIVERSITY PRESS

CAMBRIDGE

NEW YORK   NEW ROCHELLE   MELBOURNE

SYDNEY

Published by the Press Syndicate of the University of Cambridge
The Pitt Building, Trumpington Street, Cambridge CB2 1RP
32 East 57th Street, New York, NY 10022, USA
10 Stamford Road, Oakleigh, Melbourne 3166, Australia

First published 1987

Printed in Great Britain at The Bath Press, Avon

British Library cataloguing in publication data

Upton, Martin
African farm management.
1. Farm management — Africa
I. Title
630'.68   S562.A35

Library of Congress cataloguing in publication data

Upton, Martin.
African farm management.

Includes index.
1. Farm management—Africa. I. Title.
S562.A34U65 1987   630'.68   86-33400

ISBN 0 521 32842 X hard covers
ISBN 0 521 33805 O paperback

PN

Contents

Preface

This book started as a new edition of *Farm Management in Africa* originally published in 1973 by Oxford University Press. However, it soon became apparent that substantial rewriting was needed in view of new developments in Farming Systems Analysis, Farming Systems Research, Natural Resource Economics, Household Economics, Investment Appraisal and the Analysis of Risky Decisions, to name a few. None the less I have retained the original structure of the book, which is divided into four parts; I Farm production economics, II Resource use, III Field investigations and IV Farm planning. It still seems appropriate to group the material under these headings and to order them in this way.

Clearly the emphasis is rather different from that of traditional farm management texts of temperate regions which generally concentrate on methods of accounting and record keeping. These techniques are important for commercial farmers of both temperate and tropical regions, but for the majority of semi-commercial, semi-subsistence smallholders they are of limited relevance. In their place, farming systems research and the associated farm survey and analysis methods are the main tools of field investigation. Such field investigations naturally lead to farm planning for improved productivity and household incomes.

The application of these tools for investigation and planning should be guided by an understanding of how household and farm management decisions are made and how farming systems develop. Although no theoretical analysis can provide a comprehensive explanation, a combination of the economic theories of production and consumption, described in the early chapters can, I believe, provide some illumination of the processes at work in farm household decision-making.

Some have argued that these are alien theories of limited relevance to African farmers, but I would dispute this. Although theories of economic decision-making have their limitations, they are of some relevance in any culture and society anywhere in the world. Furthermore, these are no longer ideas introduced to Africa from outside. Much of my farm economics has been learned from African friends and colleagues, particularly those to whom I dedicate this book:

THE LATE PROFESSOR H. A. OLUWASANMI
PROFESSOR Q. B. O. ANTHONIO
PROFESSOR R. O. ADEGBOYE
PROFESSOR S. LA-ANYANE

Acknowledgements

Several colleagues at Reading University have commented on drafts of parts of this text. In particular I would like to thank Professors Hugh Bunting, and Colin Spedding, Drs Doug Thornton, James Roger and Tony Woods and Steve Wiggins. They are not responsible for any errors or omissions which remain.

For speedy and accurate word processing I must thank Audrey Collins, Maxine Oakey and Stephanie Cave, and for the artwork I thank Yalemberhan Kebede. For much patience with a more than usually distracted husband and father I thank Veronica, Tim, Stephen, Rebecca and Ben.

PART I

Farm production economics

1

Farming systems and their management

The role of agriculture in African development

Agriculture is the most important industry of tropical Africa. It employs a majority of the working population and generally contributes between 20 and 60 per cent of the value of total production, as measured by Gross Domestic Product (see Table 1.1). Hence economic development, growth of Gross Domestic Product and the welfare of the majority of African people are heavily dependent on the performance of the agricultural industry. The very poor growth or stagnation of agriculture is therefore rightly seen as a crisis.

The most disturbing aspect is the steady decline in food production per head of population, which has taken place over the last fifteen years. Generally the growth in food production of about 1.3 per cent per year for sub-Saharan Africa has been insufficient to keep pace with population growth of over 2.5 per cent for the region. As a result it is estimated that food consumption per person has been falling, so that many people have a seriously inadequate diet. Several governments have resorted to imports on a large scale, particularly of wheat and other staple grains, but also of sugar and dairy products. Overall, this means that Africa has changed from a position of self-sufficiency in the basic necessities of life to growing dependence on overseas producers (some trends are illustrated in Table 1.1). Expansion of food imports has contributed to growing balance of payments deficits and external public debts, while food prices have risen or, where food prices are controlled, the costs of subsidies have increased. These trends are likely to cause inflation and generally put a brake on economic growth.

Meanwhile the production of cash crops for export has stagnated or even declined. For many major export commodities, listed in Table 1.2, Africa's share of world trade has fallen substantially over the last ten years. Thus food and cash crop and livestock production have stagnated together, so there is no support for the view that declining food production has been caused by switching resources into export crops. Nor can it be claimed that the poor export performance is due to falling world prices, since the unit value of African agricultural exports rose faster over the ten-year period to 1984 than did the unit value of exports worldwide.

African governments are increasingly concerned to bring about an agricultural revolution which will reverse these adverse trends in production. The challenge is to double the rate of growth of agricultural production within the coming decades. However, although governments may facilitate, stimulate and promote the growth of agricultural production, success must ultimately depend on the decisions of the multitude of farm households making up the agricultural industry. In short, agricultural development must occur at the farm level.

Furthermore, in most of Africa, rural people make up the majority of the poor and disadvantaged. Many of the urban unemployed have recently migrated from their villages in the hope of improving their welfare. Hence increased crop and livestock production is needed to raise farm incomes, improve the level of living and reduce the rate of out-migration from rural areas.

Even where co-operative, collective or state farms are introduced the success of the policy still depends upon decisions made by managers at the farm level. It is therefore important that everyone concerned to promote agricultural development should understand something of farm level decision-making; of the problems that are faced and the way choices are made. This is the subject matter of this book.

Table 1.1. *African economies, agriculture and food production*

	Low income economies in Africa	Middle income economies in Africa
Annual population growth (1973–1984) per cent	2.9	3.0
Population per sq km (1984)	16.5	23.8
Agriculture's contribution to GDP (1984) per cent	39	25
Percentage of labour force in agriculture (1981)	78	60
Annual growth of Gross Domestic Product (1973–1984) per cent	2.0	1.6
Annual growth of Agricultural GDP (1973–1984) per cent	1.4	0.1
Index of food production per capita (1974–1976 = 100)	92	92
Cereal imports per capita (1984) kg	20.2	32.7
Annual growth rate 1974–1984 per cent	7.3	13.5
Food aid per capita (1984) kg	8.1	3.4
Annual growth rate 1974–1984 per cent	10.1	16.0

From World Bank (1986) *World Development Report 1986*, Oxford University Press.

Table 1.2. *Changes in agricultural exports 1974–1984*

	Annual change (per cent)		African exports as percentage of world total	
	Africa	The world	1974	1984
Quantity in tonnes of:				
Cocoa	−0.9	+0.9	74.7	62.8
Coffee	−2.5	+2.2	35.0	21.8
Groundnuts	−13.8	−2.2	45.4	12.8
Groundnut Oil	−2.3	−0.6	49.6	41.9
Palm Oil	−9.6	+9.9	12.0	1.7
Bananas	−7.3	+0.4	6.9	3.1
Maize	−21.5	+3.1	6.5	0.4
Rubber	−1.5	+1.1	6.0	4.6
Sisal	−10.1	−8.7	61.8	48.5
Cattle	−1.7	+1.7	24.3	17.3
Sheep and goats	+1.4	+7.0	33.0	19.2
Value of all agricultural products	+1.4	+6.5	8.1	5.0
Unit value – price index	+4.5	+3.0	—	—

From Food and Agriculture Organization (1974 and 1984) *FAO Trade Yearbook*, **28** and **38**, Rome, FAO.

The role of farm management

Farming in much of Africa is a family activity, which means that decisions about the management of the farm are closely linked with household decisions on what to eat or how to spend the time. The typical unit of production is a nuclear family comprising a man, his wife or wives and their unmarried children, although other relatives may be involved. Such a family may live in more than one dwelling, but all members generally 'share the same pot'. In many parts of Africa, female-headed farm households are common especially where there are employment opportunities for men in off-farm work. Agricultural pluri-activity is very common, and it is suggested that 25 to 30 per cent of the annual labour supply of rural households in sub-Saharan Africa is spent in off-farm activities (see Eicher & Baker, 1982). Remittances from family members living away also contribute to family income, and household decisions relate to both on-farm and off-farm activities (see Chapter 7).

Given that the family is the production unit, a farm may be defined as all the agricultural activities under the control of a farm family. Used in this way the word 'farm' means all the plots of land cultivated by the family which of course differs from the common idea that a 'farm' is a single plot. We also include pastoralist families with the resources under their control in this broad definition. Within the family decision-making is shared in various ways. In some cases, decisions may be made jointly by all the family members, while in other cases there may be division of responsibility, for instance where subsistence crops are produced by women, while cash crops are tended by the men. Frequently individual members of the household may have independent control of some plots of land or groups of livestock.

The productive resources under the managerial control of the family are conveniently classified under the headings of land, labour, and capital. A typical African smallholder cultivates a relatively small area of land, but frequently there is a larger area under his control which may be fallow or grazed by livestock. In fact the term 'land' is generally assumed to include all 'natural resources' such as water, minerals, natural-grazing, forests and other wild flora and fauna. All these are important to the African farm household. The labour-force is mainly made up of family members, although some work may be done communally and some labour is hired. Capital consists of everything else used in production, which is not a gift of nature but which has been produced in the past. It is frequently claimed that African smallholders use relatively little capital per person employed, in comparison with large commercial schemes. Although this may be true in total cash value terms, the smallholder uses many kinds of capital including permanent crops, livestock, tools, equipment, buildings, land improvements and stocks of seed, fertilizer, animal feeds and agro-chemicals. The cocoa, coffee, tea or rubber producer has a large amount of capital tied up in his trees, while a pastoralist similarly has a large amount represented by his herds and flocks.

Although we have emphasized the decision-making aspect of management, there are other tasks involved in organizing and operating a farm. Even though farmers may not analyse their management activities, they must in fact (i) set objectives, or have some view of what they are trying to achieve (ii) make decisions as to what to produce, what methods to use, how much to produce and where and when to sell (iii) implement these decisions by organizing and allocating resources and (iv) control the operation of the farm. These managerial tasks are not just done once but must be carried on all the time in the light of changing circumstances.

The study of farm management is intended to improve our understanding of these processes. Any programmes or policies aimed at accelerating the development of small farms should be based on an awareness of their relevance and likely impact. For instance price policies to provide production incentives should take full account of costs of production and farmers' price responsiveness. Programmes for

institutional change, aimed at making resources and new knowledge more readily available to farmers are doomed to failure if they are not guided by an understanding of existing institutions, current practices and farmers' attitudes. Above all, research and development of new technology should be directed towards meeting farmers' objectives and overcoming their constraints besides building on the indigenous technology already in use. Clearly this depends upon a knowledge of what farmers do and why they do it.

Such knowledge can only be obtained from the farmers themselves, so field studies and data collection are essential. Part III of this book is concerned with this topic. However, analysis of the findings requires a theoretical framework, which may be provided by the economic theories of production discussed in the remainder of Part I and resource use, dealt with in Part II. Finally Part IV describes some approaches to farm planning and evaluation of the impact of new technologies or policies.

Farming systems

A useful method of analysis is to think of a farm as a system, or set of interrelated components. The components are the resources described above, the productive activities with their associated input–output relationships, and, in family farming, consumption and other household activities. Any system has a boundary, which separates it from the larger systems which make up the environment. A farming system is under managerial control and the boundary represents the limits of that control.

Farming systems are complex; firstly because there are many fundamentally different components such as plants, animals, people, tools and wells, which means that a multidisciplinary approach is needed to study them, and secondly there are many different ways in which the components can be combined. The disciplines needed obviously include crop and livestock sciences, but may also include soil science, hydrology, engineering, economics and sociology. However, to find our way through the complexity and increase our understanding we must simplify. This is achieved *either* by concentrating on particular subsystems, such as the cereal growing enterprise or an individual animal, or preferably, by ignoring many

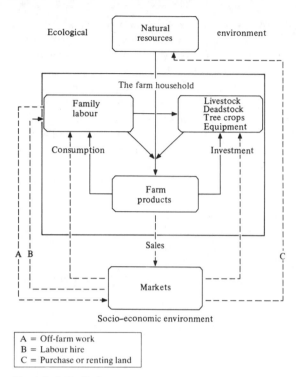

A = Off-farm work
B = Labour hire
C = Purchase or renting land

Fig. 1.1 A family farm system

inter-relationships and concentrating on a few critical ones. Such a simplified, theoretical model of a farm may be represented diagrammatically as in Figure 1.1.

This diagram serves to illustrate that inputs of labour, natural resources and capital services are combined to yield farm products, and that these products may be consumed or invested. Investment simply means additions to the stock of capital, for instance when animals are retained for breeding, or yams are kept for planting sets. Other forms of capital, such as land improvements are created directly using farm labour, as shown by the arrow from labour to capital. The broken lines represent additional relationships for farmers engaged in the cash economy, which is the case throughout most of Africa today. Then some products are sold and the proceeds may be used to buy consumer goods and items of capital, such as tools and machinery, or to hire labour and capital services or even land. Alternative off-farm uses of labour and capital such as investment in education are also represented in the dia-

gram. The farming environment consists of larger systems, of which the farm is only a part. Farming systems are particularly influenced by (i) the ecological or natural and (ii) the socio–economic environment.

The ecological environment

Human societies have always exploited the natural environment to meet the necessities of life. In doing so the balance of nature, or ecological equilibrium, is altered. Hunters and gatherers, by harvesting certain species of wild animals and plants, reduce the population, or biomass, of these species. By grazing the natural rangelands, pastoralists' herds may displace some wild game, and change the pasture composition. Cultivators modify the environment more drastically by clearing the natural vegetation, and growing new kinds of plants. Under irrigation, the amount of water available for plant growth is increased, thereby transforming the environment and allowing the introduction of exotic crops. Thus agriculture creates new man-made environments or agricultural ecosystems.

Many ecosystems can survive in a relatively stable and unchanging equilibrium, although there is always some climatic variation between seasons and between years. This may be true of pastoralism at low livestock densities, shifting cultivation with long fallows or indeed well-managed irrigation schemes, as well as virgin bush. However, if the balance is seriously disturbed, irreversible changes may occur. Careful environmental management is a necessary element of farm management to conserve natural resources and avoid their degradation. An alternative view is that the environment is constantly changing through natural evolutionary processes and under the impact of human activity. According to this view, farming systems must also change and adapt to the changing circumstances.

Despite man's ability to modify the natural environment to some extent, he is none the less constrained in his choice of crop and livestock enterprises and methods of production by the climate, soils and biology of his habitat. These factors determine the ecological zones illustrated in Figure 1.2. In tropical Africa, the most important climatic constraint is rainfall, since sufficient solar energy for crop growth

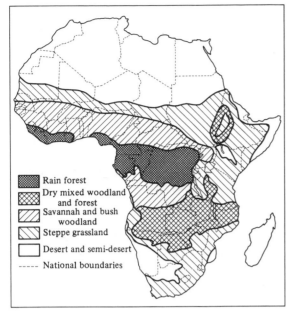

Fig. 1.2　Major vegetation zones of Africa

Rain forest
Dry mixed woodland and forest
Savannah and bush woodland
Steppe grassland
Desert and semi-desert
----- National boundaries

reaches the land surface throughout most of the year. The highlands of East Africa are exceptional in being cooler as a result of the altitude. Temperate crops such as wheat and barley may be grown, while intensive dairy farming raises fewer problems than in the hotter tropical lowlands. Bananas and similar fruits are staple foods in some highland (as well as some lowland) areas, while tea, coffee and pyrethrum are also grown. In most of Africa the rainfall regime is characterized by distinct wet and dry seasons. During the dry seasons, soil moisture loss through evapotranspiration exceeds the amount reaching the soil, so plant growth generally ceases. Thus the vegetation and the crops which may be grown are strongly influenced by the length of the rainy season.

The humid, tropical rain-forest zone of Equatorial Central Africa and the Guinea-Coast of West Africa is potentially the most productive since there is sufficient rainfall for crop growth throughout most of the year. Even where dry seasons do occur, towards the fringes of this zone, there are two rainy seasons (bimodal rainfall distribution) with a combined length of seven months or more so that normally two crops can be taken each year, while tree crops like cocoa and oil-palm flourish. Although rice is

important in some countries and maize is widely grown the basic staples are generally root crops like yams and cassava. The sub-humid Savannah zone lying outside, both north and south of the humid forest zone is drier with a single rainy season of between three and seven months' duration. A wide range of crops may be grown, although the main cash crops are generally cotton, groundnuts or tobacco, while cereals are grown for food. The drier Sahelian zone further north, which extends across the continent from Senegal to the Sudan, is similar to those parts of Southern Africa (in Angola, Namibia, Botswana, Zimbabwe and Mozambique) where growing seasons are of less than three months' duration on average. Without irrigation, cropping must be based on short-growing season crops like sorghum and millets and may be impossible in some years. Much of the area is under natural rangeland and utilized by transhumant pastoralists. In this zone in particular rainfall variation is a major cause of uncertainty. Droughts like those of the 1970s and 1980s can prove catastrophic for crop and livestock production.

Soil conditions and plant growth

Soil conditions also act as a constraint on agricultural production and careful management is needed to maintain soil structure and fertility. However, soil conditions vary with the landscape topography over quite small distances. This is one reason for the diversity of local ecological conditions which farmers have learned to exploit. Of particular importance are the swamps and valley bottoms with relatively fertile and well-watered alluvial soils. These may be suited to the cultivation of swamp rice, sugar cane and various vegetables.

Most African soils are low in organic matter, since it is rapidly broken down by micro-organisms in the warm, moist conditions at the start of the rains. Essential plant nutrients such as phosphates and nitrates are thereby released and leached out of the top soil quite rapidly. This has important managerial implications, first that crops must be planted early at the start of the rains to benefit from the flush of nutrients and second that vegetative matter must be returned to the soil to provide for the next season. Since all forms of agriculture remove essential nutrients from the system in the harvested biomass, these must be restored by some means for the system to continue. Under shifting cultivation, nutrients are returned to the topsoil during the bush-fallow period as a result of the accumulation of vegetative matter. Since trees and shrubs are deeper rooting than grasses they raise nutrients from lower levels, so that regeneration of fertility is much more rapid under forest fallow than under grass fallow. In forest areas, a suitable combination of trees with fertilizers, zero tillage and mulch farming can provide a stable system of continual cultivation. In drier regions, where deep-rooting crops are not available, grass fallows or leys are needed in the rotation, while inputs of manures or fertilizers must be provided from outside the system.

Another characteristic of low organic-matter soils, is that they are easily eroded under intense tropical rainfall when bare of vegetation. Thus it is important to maintain plant cover to minimize the possibility of erosion and desertification. Traditional practices such as mixed and sequential cropping help to ensure continuous plant cover of the soil. The introduction of continuous monocropping, in parts of the Sudan for instance, has lead to dust-bowl conditions and loss of top-soil. In fact, many cultivation and livestock husbandry practices of the African farmer are rational adaptations to the natural environment. There are serious dangers in abandoning these practices without exploring the environmental impact.

The natural vegetation and biological environment is itself influenced by climate and soils but it too affects systems of land use. Apart from the importance of plant cover and soil organic matter in maintaining fertility, the vegetation determines whether grazing or browsing livestock are more appropriate. The presence of trees in forest zones may preclude the introduction of the ox plough or tractors in view of the difficulties and costs of destumping. Thus the techniques of production are influenced by the type of vegetation.

Although the natural vegetation is modified by cultivators and pastoralists, wildlife is still exploited in various ways. Some hunting and gathering of wild fruits is widely practised, while wild oil-palm fruit and palm-wine are harvested in the rain-forest zone. Similarly wild Acacia trees are tapped for gum in more arid regions. However, the most widely consumed, naturally occurring plant material in Africa is undoubtedly firewood, which is the main fuel used in almost all rural areas. It is a renewable resource, but regrowth of timber is slow. Where consumption rate exceeds the rate of regrowth, not only are there possible adverse effects on plant cover and soil fertility, but also increasing costs are incurred in travelling further afield to collect firewood. Here, then, is another case for conservation of certain elements of the natural environment.

The other elements of the biological environment which affect agriculture are weeds, and crop or livestock pests and diseases. Unfortunately, where climatic conditions are favourable for plant and animal growth, crop competitors and parasites also flourish. Thus competition from weeds, insects and diseases is generally most severe in the humid zone where crop growing seasons are longest. Crops such as cotton may be excluded from this zone, because of pests and disease while cattle are virtually excluded by the existence of *surra* or *ngana* (trypanosomiasis) sickness carried by the ubiquitous *tsetse fly* in humid areas. Thus the incidence and spread of pests and diseases constrain land-use systems, while traditional practices such as bush burning to destroy weeds, or transhumant pastoralism to avoid the tsetse fly in the wet season, are adaptations to them. Climatic variation together with the unpredictable incidence of pest and disease outbreaks are major sources of environmental risks for the African farmer.

Population supporting capacity

Africa is often portrayed as an underpopulated region with vast areas of unused land. Certainly average population density over the whole continent is low by international standards; less than one-tenth of that of Monsoon Asia. However, it is wrong to assume that all the unused land is available for extending the area under cultivation. Not only are large areas unsuited to agricultural development because of environmental factors such as extreme aridity or prevalence of serious diseases, but also land under bush-fallow is a part of the existing cultivation system and therefore unavailable for expansion. In fact a recent FAO study suggests that Nigeria, Kenya, Ethiopia and several Sahelian states are already in a situation where complete food self-sufficiency is impossible without a change to more 'intensive' systems of land use (Higgins *et al.*, 1982).

The *intensity* of land use is measured by the amount of effort expended per hectare and the corresponding productivity which, in turn, determines the carrying capacity. One of the least intensive, or more extensive, forms of land use is pastoralism, which may support only one family or less per square kilometre. Cultivation systems are more intensive, but shifting cultivation with long forest fallows of, say, twenty years following two years of cultivation results in only one-eleventh of the available area being used at any one time. Hence productivity per hectare is only one-eleventh of the yield per cropped hectare. Shorter bush or grass fallows allow more intensive land use and therefore a higher population carrying capacity, but maintenance of soil fertility requires more effort and possibly inputs of plant nutrients from outside the system. Continuous cropping is yet more intensive, but can only be sustained by the use of manures and fertilizers or practising forms of agroforestry mentioned earlier.

Thus human population density may be a causal factor influencing intensity of land use. Population growth apparently causes the spread of cultivation at the expense of pastoralism, the shortening of bushfallows and a general intensification of land use. However, the above discussion should have made it clear that man's choice of land use systems is limited by his environment. For instance, irrigation, which is probably the most intensive form of agriculture, is impossible where water is not available. In fact some migration occurs from areas of low potential to areas of high potential, so populations may grow faster in the latter situations. None the less, political boundaries and social constraints limit migratory movements, with the result that, while parts of the Sahelian zone of low potential are possibly overpopulated, much of the central African rain forest with high production potential is very sparsely populated.

For the continent as a whole, and for the main sub-regions of West Africa, Eastern and Central Africa and Southern Africa, there are sufficient natural resources to meet the food needs of rapidly growing populations and yield surpluses for export. There is considerable potential for growth in agricultural production. To explain the current stagnation of agriculture, therefore, we must turn to the constraints imposed on farming families by the socio–economic environment.

The changing socio–economic environment

There is a widespread, though probably idealized, view of traditional African society as being made up of small, self-sufficient, communities of people linked through kinship ties. Production, distribution and consumption all take place within the 'closed' community while social and economic relations are based on the status of individuals as members of the lineage. Thus help is offered to a neighbour because he is a kinsman or a senior member of the society. Impersonal contractual relationships between comparative strangers are comparatively unknown. The allocation of natural resources and labour are the result of a system of reciprocity or sharing combined with some redistribution under the central authority of the chief or elders of the lineage. There are few private possessions other than livestock and limited accumulation of wealth. The degree of inequality of income and power, between leaders and others, or between men and women, varies from one cultural group to another. None the less, considerable support for the weak and disadvantaged is offered through the extended family system.

Traditional society is also typified by strict observance of social norms and customs, including religious beliefs and practices. The allocation and inheritance of land is governed by customary rules and so too is the provision of communal labour for specific tasks. Agricultural practices such as mixed cropping or the making of heaps for certain crops, though originally based on sound principles, may become embodied in the traditional culture. The Muslim religion, long established in much of the Sahel and East Africa, influences farming systems through such observances as purdah, which limits the availability of women for fieldwork, and the prohibition of pig meat.

These established customs, laws and relationships make up the traditional institutions which govern the allocation and use of resources and the distribution of agricultural products. However, it is dangerous to generalize. Africa has a greater diversity of peoples and cultures than any other continent. Furthermore, trade and market exchange reached Africa many centuries ago and today there are very few traditional, self-sufficient, subsistence farming communities left.

Although trade had much earlier beginnings, the rapid growth of the market for African export crops during the last century and a half radically altered the socio–economic environment and had far-reaching effects on the evolution of farming systems. Many countries became, and have remained, heavily dependent on exports of one or two cash crops for their foreign exchange earnings. For example, Uganda depends heavily on coffee, cotton, and tea and the Sudan on cotton. For farmers it led to diversification into new crops and increased production of commodities for sale. Export-crop producers now had more money to spend on farm inputs and consumer goods, so trade in these items expanded in rural areas. Regional agricultural specialization increased with those regions unsuited to export-crops concentrating on producing a marketed surplus of food for sale to export-crop growers as well as to those engaged in non-agricultural employment. Thus trade and commerce spread through most rural areas; marketing became an important call on the time of farm families while the numbers of specialist traders and produce-buyers increased. Transactions based on impersonal exchanges, for money, replaced some of the traditional relationships based on kinship and sharing.

The increased commercialization of agriculture was accompanied, in some countries at least, by the accumulation of capital and social differentiation between 'commercial' or capitalist farmers, 'traditional' farmers and hired workers. This was most marked during the colonial era of settler farming in East and Central Africa, where governments are still faced with problems of designing appropriate policies for the two very different types of farm system. Elsewhere, the establishment of permanent crops such as cocoa, coffee and tea represented a significant investment by farmers. Families which laid claim to wild-growing trees such as oil-palms and rubber thereby accumulated capital assets. Labour hire for wages expanded and prompted large-scale migra-

tions such as that from Upper Volta, Mali and Niger to the groundnut exporting regions of Senegambia and the cocoa-belt of the Ivory Coast, Ghana and Nigeria.

Towns and cities are important features of the socio–economic environment, not least because they are the main centres of commerce and political power. It is estimated that Africa's urban population is growing at over six per cent each year and now represents about one-fifth of the total. In spite of the high rate of rural–urban migration reflected in these figures, rural populations are still growing in absolute numbers. None the less, urban populations are growing much faster as an inevitable part of the development process, since manufacturing industry requires large concentrations of workers. The scope for specialization, division of labour, and exchange is greatly increased as towns grow. Although some commentators have emphasized the sharp differences that exist between town and country, or centre and periphery, there are important links between the two. First, those who migrate to town usually retain their links with the village, may send remittances home and may eventually return there in old age. Second, through the formal education system, agricultural extension and mass communication, there is a flow of knowledge and information mainly from centre to periphery. An improvement of the flow in the opposite direction from farmers to planners, administrators and scientists is highly desirable. Third is the need for an increasing marketed surplus of food for the urban population and raw materials for urban industry, produced from rural areas. Finally, there is a contrary flow of agricultural inputs and consumer goods from the towns to rural areas.

Markets

For all traded commodities and inputs, markets represent the interface between the household and the wider socio–economic environment. The world 'market' refers here to all the financial transactions for a particular commodity, over a given time period in a given region and not just to a market-place. The farmer receives 'signals' from the market regarding consumer requirements in the form of prices. The relative prices of outputs to inputs (also known as the 'terms of trade') are important in determining both the income and welfare of the farm family *and* what and how much is produced. The 'income effect' is obvious in that an increase in the product price (or an improvement in the terms of trade) must yield a larger total cash income, even if there is no change in the farm system. However, an increase in the price of a particular product is likely to encourage farmers to produce more of that product for sale. The price rise may induce farm families to work harder and produce a larger marketed output in total, although some theorists have predicted the opposite effect (see Chapter 7 for fuller discussion).

Apart from this there is an incentive to transfer resources from some other activity into producing more of the higher priced crop. Thus, when in the late 1970s the price of sunflower seed rose relative to that of maize in Zambia, farmers switched some of the area previously under maize into sunflower production. In fact, there is extensive evidence from all over Africa of a positive supply response to crop prices. (For a summary of the evidence see Askari, H. & Cummings, J., 1976.) An obvious example is the rapid increase in market supplies of rams every year just before the Muslim festival of *Idd al-adha*. This evidence is important in proving that smallholders are 'economic men' who respond to price incentives, even though making money is probably not their sole aim.

The price a producer receives depends upon the effective demand for his product; or, roughly speaking, the size of the market in relation to the price level. A change in the demand for an agricultural commodity might be caused by any combination of the following.

(i) Population growth leads to increased demand and prices, since the number of consumers is increased.

(ii) Rising consumer incomes are associated with increased consumer spending. Note, however, that spending on food does not normally increase as fast as total spending, while purchases of 'inferior goods' such as *gari* may actually decline.

(iii) Changes in taste in favour of a particular commodity, such as the shift to bread consumption in many African cities, cause an increase in its demand.

(iv) Changes in the supply of substitute commodities have an opposite effect on demand for the original commodity. For instance the development of synthetic fibres has undoubtedly reduced the demand for cotton, relative to what it would have been in their absence.

Export crops are sold in a world market, which is large relative to the amounts any one African country supplies. World population and *per capita* incomes are growing and, as a result, so too is market demand for most primary commodities, although at a much reduced rate in the world recession of the early 1980s. However, the demand for agricultural commodities grows less rapidly than that for most manufactured goods, partly because as incomes rise a declining proportion is spent on food and partly because of the development of synthetic substitutes for cotton, rubber and other non-food crops. As a result there is a tendency for the terms of trade to move against agricultural exports, which means that their prices fall relative to those of imports. This can occur even when prices of exports are rising, if prices of imports rise faster. It implies that a steady increase in production for export is needed in order to pay for imports and avoid growing trade deficits.

The demand for domestically marketed food crops and livestock increases with population growth, migration out of agriculture and rising incomes. Where increased imports and domestic food production have failed to keep pace with the growth in demand, food prices have risen. Although rising food prices may create social, economic and political problems among the urban populations, they do provide an incentive for farmers to increase their marketed surplus.

Prices also vary, inversely, with changes in supply as illustrated by the seasonal fall in price of most crops at the time of harvest when the supply is increased. Chance variations in climate or disease incidence are important in this respect, with marked price increases occurring in drought years, when crop yields and hence supplies are reduced. At such times, pastoralists generally try to reduce their livestock numbers, by increasing the supply of animals offered for sale. Hence cattle prices may fall during a drought.

Input prices also influence the farmer's terms of trade, and in turn, are influenced by the amounts available. Markets and supply facilities are often poorly developed for material inputs such as improved seeds, fertilizers, agro-chemicals, equipment and fuels. Supplies are limited by lack of local manufacturing capacity, import controls and problems of distribution in rural areas. Scarcities and related high and possibly fluctuating prices are serious disincentives to the use of purchased inputs. Weak delivery systems are perhaps the greatest constraint on progressive change in Africa today. The allocation of land and labour and even financial capital still takes place largely outside the market system. We will return to these issues in the appropriate chapters on resource use.

Locational effects

Since the major assembly points for export commodities, the largest retail markets for food and the supplies of material inputs to farmers are all located in towns, distance from these centres, or from their main transport routes, has an important influence on farm prices and hence farming systems. As transport costs increase with increasing distance, the prices farmers must pay for their material inputs rises and the price they receive for their marketed commodities falls; in other words their terms of trade deteriorate with increasing remoteness. Thus location influences both the intensity of land use or level of output per hectare and the choice of enterprise, since transport costs per unit of value vary between different commodities, depending on their bulk and perishability. Many African cities such as Kano and Zaria in Nigeria are surrounded by a 'close-settled' zone, where continual cultivation is supported by manures carried out from the town. Vegetables and other high value crops are produced for the urban markets. Within this peri-urban zone are frequently found specialist large-scale poultry units and dairy holdings, set up by urban business men, civil servants or academics. Sophisticated technology, exotic livestock and large amounts of capital are involved.

Beyond the close-settled zone there is a wide belt of, mainly, crop production for the market with land use intensity and product perishability declining with distance. In many countries such a belt of 'commercial' production is found along the main line of rail

and main roads. More remote areas must depend increasingly on production for their own subsistence. Livestock, however, can be moved large distances relatively cheaply on the hoof. Some of the more remote areas in the semi-arid and arid zones are devoted to pastoralism, based on milk production for subsistence and the sale of animals in order to buy cereals and other consumer goods.

Clearly the development of roads, railways and other communications, together with more widely dispersed market-places, storage facilities and processing plants will improve the terms of trade for those farmers that are reached. The problems and costs of extending delivery systems in this way are particularly acute for large, sparsely populated countries like the Sudan and the rest of the Sahel or Zambia, Botswana and Namibia in Southern Africa.

The role of government

The national government is an important component of the socio–economic environment, capable of influencing most other features, favourably or unfavourably. Even the natural environment may be changed through programmes of soil or wild life conservation or disease eradication. Views differ, between people of differing political persuasions, as to how much government intervention is desirable. As a bare minimum all governments are expected to provide certain social services such as public administration, public works or social overhead capital, law and order, defence, communications and the like. In addition most governments take responsibility for regulating foreign trade, or at least currency exchange rates, and for promoting economic growth. In all this, the aim of ensuring an equitable distribution of incomes may or may not loom large.

It should hardly need pointing out that governments also operate under constraints. Funds, from taxes, loans and aid, are limited, so governments too face problems of allocating scarce resources among competing ends. Development theories prevalent in the 1960s, when many African countries achieved their independence, emphasized the importance of industrialization. Agriculture as the largest sector was expected to release the necessary capital and labour resources. This was known as the 'factor contribution' of agriculture. The government's role in promoting development, was seen as extracting the surplus from agriculture to finance the growth of other sectors.

Although many now believe that these policies have failed and there is widespread recognition of the need and the potential for agricultural growth, governments have not made the necessary policy changes. A recent World Bank report on African economic development refers to three 'critical-policy inadequacies': a) trade and exchange-rate policies have overprotected industry, held back agriculture and absorbed much administrative capacity, b) too little attention to administrative constraints and c) consistent bias against agriculture in price, tax and exchange-rate policies.

Trade and exchange-rate policies of many countries involve the maintenance of an overvalued currency, or an artificially low price for foreign exchange, maintained by strict import and foreign currency controls. This serves to hold down prices farmers receive for export crops, and the price paid for imported cereals and other foodstuffs, for which foreign exchange is released. Thus imports are encouraged and food prices are held down to the disadvantage of domestic producers. These policies together with taxes on export crops and widespread price controls on staple foods, have a substantial adverse impact on the farmer's terms of trade. Thus, it is argued that removal of controls and a return to free market prices will provide the necessary price incentives for agricultural growth besides improving rural incomes. The report also suggests that constraints on administrative capacity can best be overcome by greater reliance on private enterprise.

However, the improvement of farm price incentives is only one of several alternative strategies for promoting agricultural growth. It is doubtful whether pricing policies alone can bring about the desired agricultural revolution. Two other strategies are deserving of government attention.

(i) *Institutional change*

This includes such policies as land reform, 'ujamaa' villagization in Tanzania, introduction of formal cooperatives or establishment of marketing boards. Most of these policies have several political objectives only one of which may be the improvement in productivity and welfare of the small farmer. However, in so far as his organization and bargaining power may be increased, so might his terms of trade. Experience has shown that careful assessment and popular assent are necessary if imposed institutional changes are to succeed economically. These requirements suggest the need for study of existing farming systems before institutional change is attempted.

(ii) *New technology*

The development and spread of new technology offers most hope of producing large increases in agricultural productivity, but Africa still awaits its 'Green Revolution'. Generally, technical knowledge is a 'public-good', theoretically available to all, so there are strong arguments for public finance of research and development, and agricultural extension. Agricultural research and development is actively pursued on government research stations in every African country besides the international institutions such as ILCA (International Livestock Centre for Africa), IITA (International Institute for Tropical Agriculture), CIMMYT (International Maize and Wheat Improvement Centre) and ICRISAT (International Centre for Research in the Semi-arid Tropics). It is increasingly realized that appropriate and acceptable new technologies will only be developed if they are based on a sound understanding of existing farming systems and local based practices.

The farmer's aims

Description of a farming system is possible simply in terms of the resources used, the crops grown, the livestock kept and the quantities produced and consumed. Analysis of the system, for diagnosis of problems and recommendation of improvements, on the other hand, requires the specification of the farmer's aims and objectives. Only then is it possible to determine which constraints are effective in preventing the attainment of objectives and how these constraints may be overcome. This presumes that farmers are rational in using their knowledge to attain given ends;

that they always have a reason for farming in one way rather than another, although the reason is not always for cash gain. Of course, different farm families, indeed different individuals within the same family, may have different aims, but, in general, most farmers pursue some combination of the following six main objectives.

(i) Securing an adequate and assured food supply;
(ii) Earning a cash income to meet other material needs;
(iii) Avoidance of risk or, more simply, survival in an uncertain environment;
(iv) Leaving time for leisure and other non-agricultural activities;
(v) Provisions for the future, for old age and the welfare of dependants;
(vi) Achievement of status and respect within the community.

Clearly the simplifying assumption, often made by economists, that the producer's *sole* aim is to maximize profit is false. So, too, is the view that his *sole* aim is survival. There are several objectives that must be reconciled in some way, by farm families and by those of us who are concerned to understand how farming systems are managed. Choices under alternative objectives will be discussed in more detail in the chapters which follow, particularly Chapters 4 and 5.

Initially we emphasize the objective of earning a cash income and indeed maximizing the cash profit. There are three main arguments to justify this initial emphasis. First, there is little doubt that this is *one* of the farmer's objectives and that it is growing in importance with increasing penetration of the market economy into rural areas. Second, the other aims might be satisfied indirectly by maximizing cash income. Provision for the future and status and respect may be achieved by maximizing cash income and accumulating wealth. Although risk avoidance usually entails giving up some profit on average, accumulating wealth is one way of providing insurance against disasters and ensuring survival.

The third justification stems from the idea that objectives may be viewed as targets or constraints.

Thus the objective of securing an adequate food supply may be viewed in this way. The farmer's choices are limited or constrained by his need to secure the target amount of food. Once that target is assured, he is free to pursue other objectives. Consideration of the list of objectives given above suggests that any or all of them could be viewed as constraints. A farmer who has ensured that the constraints represented by his needs for security, leisure and status can be satisfied may then aim to maximize his residual money profit. The emphasis on this objective is justified on the grounds, not that this is the farmer's primary objective but rather, that other, higher priority objectives have already been allowed for.

Further reading

Anthony, K., Johnston, B., Jones, W. & Uchendu, V. (1979). *Agricultural Change in Tropical Africa*, Ithaca, Cornell University Press
Askari, H. & Cummings, J. (1976). *Agricultural Supply Response: A Survey of the Econometric Evidence*, New York, Praeger
Chambers, R. (1983). *Rural Development: Putting the Last First*, London, Longmans
ECA (1976). *Survey of Economic and Social Conditions in Africa*, Addis Ababa, UN Economic Commission for Africa
Eicher, G. K. & Baker, D. C. (1982). *Research on Agricultural Development in Sub-Saharan Africa: a Critical Survey*, Michigan State University, East Lansing. International Development Paper 1
FAO (1981). *Food Plan for Africa*, Rome, Food and Agriculture Organization
Higgins, G. M. *et al.* (1982). *Potential Population Supporting Capacities of Lands in the Developing World*, Rome, FAO
La-Anyane, S. (1985). *Economics of Agricultural Development in Tropical Africa*, Chichester, Wiley
Levi, J. & Havinden, M. (1982). *Economics of African Agriculture*, London, Longmans
Richards, P. (1985). *Indigenous Agricultural Revolution: Ecology and Food Production in West Africa*, London, Hutchinson
Ruthenberg, H. (1980). *Farming Systems in the Tropics*, 3rd edn, Oxford University Press
World Bank (1981). *Accelerated Development in Sub-Saharan Africa: An Agenda for Action*, Washington, IBRD

2

The production function and economic optimization

The input–output relationship

In our discussion so far we have simply assumed that, within a farm system, inputs of natural resources, labour and capital are converted into outputs of food and other commodities. To proceed further we now suppose there to be some measurable relationship between the quantities of inputs used and the quantity of output produced. Such a relationship is known as a 'production function'. There are many practical and theoretical problems involved in estimating a production function, and these will be discussed in some detail in Chapter 13. Nonetheless it seems reasonable to assume that the farmer has some idea of the amount of yield he can expect, on average, when he decides to allocate inputs to, say, growing a plot of maize. He therefore has some knowledge of his production function. Of course, he cannot predict the yield he will obtain with certainty but he knows what to expect, on average, from a given combination of inputs and bases his decisions on this knowledge.

The simplest kind of farm management decision, which will now be analysed, concerns the allocation of a single variable input, in this case hired labour for weeding, to a single product, in this case maize. Thus, we assume that the farmer knows what yield he can expect, on average, for different amounts of weeding effort. This knowledge is based on past experience on his own and other farms. The figures given in the first two columns of Table 2.1 serve to illustrate the relationship which might exist between the number of days spent weeding a maize plot and the yield obtained. The important point to note about these figures is that the yield, or total product, increases with increased labour use but at a diminishing rate.

The effect is seen rather more clearly, when we

Table 2.1. *The effect of varying weeding effort on maize yield*

Days of weeding labour	Maize yield (Total product)	Average product	Marginal product[a]
	Bags of maize		
0	0	—	
			2.5
1	2.5	2.5	
			3.5
2	6.0	3.0	
			2.0
3	8.0	2.67	
			1.2
4	9.2	2.3	
			0.7
5	9.9	1.98	
			0.3
6	10.2	1.70	
			0
7	10.2	1.46	

[a] The marginal product is calculated as the difference between successive values in the total product column.

calculate the average product (AP equals total product divided by the number of man–days worked) and the marginal product (MP equals the *increase* in total product for each additional day of weeding labour) as shown in columns 3 and 4 of Table 2.1. Clearly, both average and marginal products diminish as weeding labour is increased. These effects are also shown in Figure 2.1 which is based on the information given in the Table. The shaded blocks in the diagram represent the marginal products of each additional man–day of weeding labour. It may be noted for any particular level of weeding labour use the total product is equal to the cumulative sum of the marginal products. For instance, when three days are spent on weeding, the total product is

$$2.5 + 3.5 + 2.0 = 8.0 \text{ bags}$$

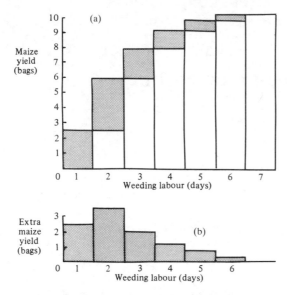

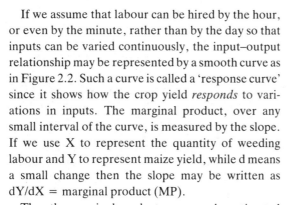

Fig. 2.1 Diminishing marginal returns to weeding labour, (a) Total product, (b) Marginal product

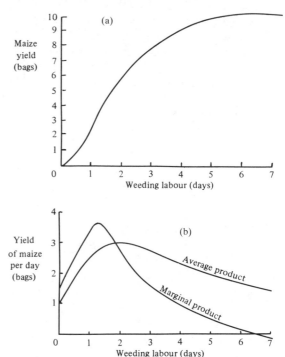

Fig. 2.2 The response curve for weeding labour, (a) Total product, (b) Marginal and average product

If we assume that labour can be hired by the hour, or even by the minute, rather than by the day so that inputs can be varied continuously, the input–output relationship may be represented by a smooth curve as in Figure 2.2. Such a curve is called a 'response curve' since it shows how the crop yield *responds* to variations in inputs. The marginal product, over any small interval of the curve, is measured by the slope. If we use X to represent the quantity of weeding labour and Y to represent maize yield, while d means a small change then the slope may be written as dY/dX = marginal product (MP).

Thus the marginal product curve can be estimated as shown in the lower diagram. Now the total product for any particular level of weeding labour use is equal to the area under the marginal product curve, up to that level of input. For example, the area under the curve when 6.5 days of labour are used represents 10.2 bags of maize, which equals the total product.

The assumption that all inputs other than weeding labour, for instance, the area of land and the amount of seed, are fixed is important. It means there are limits on the amount of maize that can be produced from this particular plot, and that is why extra weeding labour becomes less and less effective in

raising the yield. This case of diminishing returns to weeding labour in maize production is an example of a general rule, the so-called 'law of diminishing returns', which we can now state precisely. *If one input is varied, the amounts of all other inputs being held constant, the marginal product per unit of the variable input eventually diminishes.* This 'Law' applies very widely and explains why we cannot produce all the food we need from a single plot. It is likely to apply as farming population pressure increases on a fixed area of land. As the labour input rises, the marginal product per person is likely to fall, unless new, more-productive systems of farming can be found.

Generally speaking the average product, often referred to as the 'productivity', of a variable input diminishes along with the marginal product when inputs are increased, as shown in Table 2.1 and Figure 2.2. However this only occurs where the marginal product is less than the average product of labour. Initially the marginal product is greater than the average product and pulls the average up so productivity rises.

The point where total product is at a maximum and marginal product is zero, at about 6.5 days of weeding labour in this case, is sometimes known as the technical optimum or technically the best choice. In this case it represents the highest attainable yield per hectare of land. However, this is to ignore the cost of weeding labour. If, in fact, labour for weeding is scarce and costly while land is freely available it might be better to maximize the productivity of labour, which from these figures implies less than two days of weeding effort.

The economic optimum

For simplicity in this first example we assume that labour can be hired at a wage rate of £3 per day, while maize can be sold for £4 per bag. This means that each day of labour costs the equivalent of 3/4 = 0.75 bags of grain to hire, or one day of weeding labour exchanges for three-quarters of a bag of grain. It is now relatively straightforward to calculate the economic optimum, where the surplus over the variable cost is maximized. Since all inputs except weeding labour are supposed to be fixed, their cost is constant regardless of the amount of weeding labour used. As a result, when the surplus over variable cost is maximized so too is the farmer's profit, which is simply the surplus over the variable cost minus the fixed cost.

The economic optimum, or profit maximizing level of weeding labour, for the assumed prices, is shown in Table 2.2 to be four days. At this level the surplus over the cost of weeding labour is maximized (see column 4) although total yield is below the maximum.

An alternative approach to finding the economic optimum, is to compare the cost per unit of labour input with the marginal product earned. The economic optimum is then found where the marginal (value) product equals the unit factor cost. In this example the unit factor cost is 0.75 bags of grain, and the marginal product is closest to this value for the fourth day of weeding labour. However, if inputs can be varied continuously, the precise economic optimum can be found using the response curve as shown in Figure 2.3. This is identical with Figure 2.2

Table 2.2. *Costs and the economic optimum*

Days of weeding labour	Cost of weeding labour	Total product	Surplus over weeding cost[a]	Marginal product
	Bags of maize			
0	0	0	0	
				2.5
1	0.75	2.5	1.75	
				3.5
2	1.5	6.0	4.50	
				2.0
3	2.25	8.0	5.75	
				1.2
4	3.0	9.2	6.20	
				0.7
5	3.75	9.9	6.15	
				0.3
6	4.5	10.2	5.70	
				0
7	5.25	10.2	4.95	

[a] This is the difference between total product (third column) and cost of weeding labour (second column).

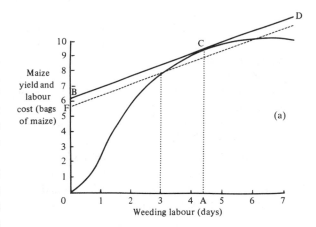

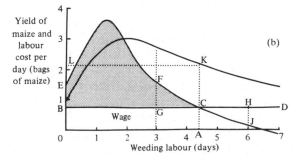

Fig. 2.3 The economic optimum input, (a) Maximum surplus, (b) Marginal product equals unit factor cost

except that the line BD has been added to represent the cost of hiring weeding labour. In the upper (total product) diagram, BD represents the total cost of weeding labour so it has a slope of 0.75 (the cost, in bags of maize per day of labour). The maximum surplus over labour cost is obtained when, as shown, the cost line just touches, or is tangent to, the response curve at C. At this point the slope of the curve is equal to that of the tangent, so marginal product equals unit factor cost of 0.75 bags ($MP = dY/dX = P/R$ where P = wage   and R = price of maize). Surplus OB is maximized at 6.25 bags, when 4.4 days of weeding labour are employed.

In the lower (marginal product) diagram, BD actually represents the unit factor cost, at 0.75 bags of maize per day. Total weeding labour cost is given by the area under this line (eg OBCA when OA days of labour are used). The marginal product is equal to the unit factor cost where the two lines cross at C, so this is the economic optimum. Since the area under the marginal product curve represents the total product, the shaded area BCE represents the surplus over weeding labour cost. This surplus would be smaller if less than OA (equals 4.4) days were spent weeding. For instance, if only three days were used, the surplus would be reduced to the area BGFE on the lower diagram, or OF (5.75 bags) in the upper diagram. If, on the other hand, more weeding labour was employed, say six days, the surplus would again be reduced. Referring back to the lower diagram, the area CHJ would be lost from the total surplus. Thus, for the given response curve and prices, the maximum surplus of 6.25 bags of maize cannot be improved upon.

Alternatively, total product may be measured as average product times total variable input. In Figure 2.3(b) this is represented by the area OAKL at the economic optimum. Thus the surplus over labour cost is represented by the area BCKL.

If the price of labour rises, or the price of maize falls, the economic optimum level of weeding falls. For example, at a labour cost of 1.5 bags of maize per day, the economic optimum weeding rate is only three days, as the reader may check from Table 2.2 or Figure 2.3. At the price of more than three bags per day it is not worth employing weeding labour at all

since this exceeds the maximum average product. Conversely, if the relative cost of labour falls, the economic optimum use of labour rises. Hence, if the farmer's objective is to maximize the surplus over weeding costs, he will employ more weeding labour if the wage falls and less if it rises. His 'demand curve' for weeding labour slopes downwards to the right.

A second variable input

If there is another variable input, such as the amount of seed sown, the level of its use will influence the economic optimum for the first input, weeding labour. An increase in seed use which raises maize yield, thereby raises the average product per day of weeding labour but may reduce the amount of weeding necessary. More specifically there may be scope for substituting seed for weeding labour in producing a particular yield of maize. A possible response-surface for these two variable inputs is represented by Table 2.3 and Figure 2.4.

The original response-curve for weeding labour is now seen to relate to a fixed seed rate of 10 000 plants per plot, as shown in the first row of Table 2.3 and by the line AB in Figure 2.4. Although some yield is obtained, even when no weeding labour is used (for seed rates above 15 000), the situation is different for seed. There can be no yield when no seed is used, so the response curves for seed pass through the origin. It is clear from Figure 2.4 that there are diminishing

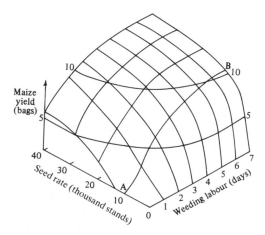

Fig. 2.4 The response surface for weeding labour and seed

Table 2.3. *Maize yield response to weeding labour and seed rate*

Weeding labour (days)	0	1	2	3	4	5	6	7
Seed rate (thousand stands)		Yield in bags of maize						
10	0	2.5	6.0	8.0	9.2	9.9	10.2	10.2
20	2.2	6.0	8.5	10.0	10.8	10.8	11.0	11.0
30	4.6	8.0	10.0	11.0	11.5	11.5	11.5	11.5
40	5.6	8.4	10.3	11.2	11.5	11.5	11.5	11.5

Table 2.4(a). *The five-bag isoquant (for five bags of maize output)*

Weeding labour (days)	Seed rate (thousand stands)	Rate of technical substitution[a]	Cost (£)
0	30		9.40
1	18	12	8.40
2	10	8	9.00
3	5	5	10.50
4	2	3	12.60
5	1	1	15.30
6	1	0	18.30
7	2	−1	21.60

[a] Calculated here as the difference between successive values in the seed-rate column.

Table 2.4(b). *The ten-bag isoquant (for ten bags of maize output)*

Weeding labour (days)	Seed rate (thousand stands)	Rate of technical substitution[a]	Cost (£)
0	∞		∞
1	60	∞	21.00
2	30	30	15.00
3	20	10	15.00
4	15	5	16.50
5	11	4	18.30
6	9	2	20.70
7	9	0	23.70

[a] Calculated here as the difference between successive values in the seed-rate column.

returns when either input is increased on its own or even when both are raised together. This is likely to occur, since other inputs, such as the area of cropped land, are limited.

Each contour line around the surface of the diagram, shows different combinations of seed and weeding labour, which will produce the same particular yield level. Contours are shown for yields of five bags and ten bags. In order to study the scope for substitution between inputs, we will now concentrate on these contours which are known as 'isoquants' meaning lines of equal output. A two dimensional plan-view is given in Figure 2.5, while data for the isoquants are presented in Table 2.4.

In the diagram (Figure 2.5), the isoquants slope

downwards from left to right. As *more* weeding labour is used, *less* seed is needed to produce a given yield, since the two inputs are assumed to be substitutes. The negative of the slope is known as the 'rate of technical substitution' (RTS) which is the reduction in seed use for each additional unit of labour input. Calculated values of the RTS, for the five-bag isoquant, are shown in the third column of Table 2.4(a), while those for ten-bag isoquant are listed in the same column of 2.4(b).

It is clear from the figures in these columns, that

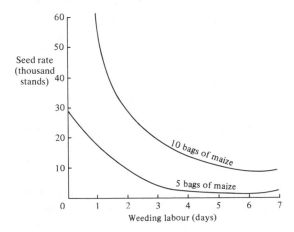

Fig. 2.5 The isoquant diagram

the RTS diminishes, as weeding labour use increases. The diminishing RTS is reflected in the fact that the isoquant curve bulges towards the origin in Figure 2.5. This effect can be explained by the law of diminishing returns. It can be proved mathematically, that the RTS is equal to the ratio of the marginal products (RTS = $-dX_2/dX_1$ = MP_1/MP_2, where d means a small change, X_1 and X_2 are quantities of weeding labour and seed respectively and MP_1 and MP_2 are their marginal products). In this case the rate at which seed use declines per unit increase in weeding labour is equal to the marginal product of weeding labour divided by that of seed. If the marginal product of labour is relatively high, one unit substitutes for a lot of seed. Conversely, if the marginal product of labour is relatively low, one unit substitutes for a small amount of seed. Hence as weeding labour use increases, with other inputs fixed and seed inputs falling, the marginal product of labour diminishes and so too does the RTS.

The least-cost combination

Given that more than one input can be varied, we may identify two aspects of the farmer's decision-making. One is how to combine the variable inputs, or what method of production to use. The second is how much to produce. We may delay discussion of the second question until the next Chapter. Here we avoid it by assuming that the farmer wishes to produce a fixed target quantity of ten bags of maize to feed his family and wants to find the cheapest way of doing so. He is seeking for the least-cost combination of variable inputs.

We now suppose that, while the wage rate per day of weeding labour is £3, the price of seed is 30p per 1000 stands. On the basis of these prices the total cost of each combination of weeding labour and seed are given in the last column of Table 2.4(b). It is clear that the least-cost combination for producing ten bags of maize is between two and three days of weeding labour and a seed rate of between 20 and 30 000 stands. For this combination the rate of technical substitution, 10, is equal to the inverse price ratio 3/0.30. In fact this is the usual method of defining the least-cost combination.

The argument may be reinforced by reference to

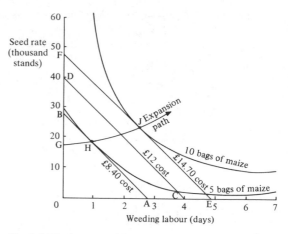

Fig. 2.6 The least-cost combination

the isoquant diagram. In Figure 2.6 equal-cost, or isocost, lines have been added. Each of these straight lines represents all combinations of weeding labour and seed which can be purchased for a given total cost. For instance, the line CD represents a total cost of £12 which would allow the hire of four days of labour at £3 per day, or the purchase of 40 000 seeds or various intermediate combinations of the two inputs. Its slope is again negative, because *more* labour can be hired from a fixed total sum, only if *less* seed is purchased. The negative of the slope is 40 000/4 = 10(000) which is also equal to the inverse price ratio, as we have already seen. The isocost line for a smaller total cost (AB) is parallel to CD but nearer to the origin, while that for a larger total cost (EF) is also parallel but further from the origin.

It should now be clear that normally the least-cost method of producing a particular level of yield is found where an isocost line is tangent to the isoquant. For a yield of ten bags of grain this occurs at point J representing 2.4 days of weeding labour and 25 000 plants with a total cost of £14.70. At a point of tangency the slope of the curve is equal to that of the tangent so the rate of technical substitution equals the inverse price ratio, which confirms our earlier conclusion.

An alternative method of defining the least-cost combination may be derived mathematically. We may recall that the rate of technical substitution of seed by weeding labour is equal to the ratio of the marginal product of weeding labour to that of seed. Therefore, since at the least-cost combination the rate of technical substitution is equal to the inverse price ratio we may write

$$RTS = MP_1/MP_2 = P_1/P_2$$

where the subscripts 1 and 2 refer to weeding labour and seed respectively. Rearranging the equation gives

$$MP_1/P_1 = MP_2/P_2$$

In words, this means that the marginal product per unit of expenditure is the same for all variable inputs. This rule for finding the least-cost combination can be extended to cover any number of variable inputs.

It should be noted that these rules do not always apply. In the case of the five-bag isoquant shown in Figure 2.6, the least-cost combination is found at point H, where very little weeding labour is used. The cheapest way of producing say four bags of maize might involve no weeding at all, at point G. For such 'corner solutions' the rules for finding the least cost combination, described above, do not apply. When no labour is used, the rate of technical substitution is less than the inverse price ratio, and the marginal product per unit of expenditure on weeding labour is less than that for seed.

The least-cost combination can be determined for any feasible level of yield. The line joining all least-cost combinations for a given set of prices (GHJ in the diagram) is known as the 'expansion path'. It traces out the combinations of variable inputs a cost-minimizing farmer would use as he increases production. This, then, is the basis for estimating the variable cost curve relating cost to total product output, to be discussed in the next Chapter.

A change in relative prices of the variable inputs alters the least-cost combination, and hence the expansion path. To illustrate this let us assume that the wage of weeding labour doubles to £6 per day. Now only half as much labour can be hired for a given sum. The downward slope of the isocost lines is therefore twice as steep, as shown by FK in Figure 2.7. It is apparent that for the ten-bag isoquant a new point of tangency occurs at point M representing a combination of one-and-a-half days of weeding labour and a seed rate of 37 000 stands. The least-cost

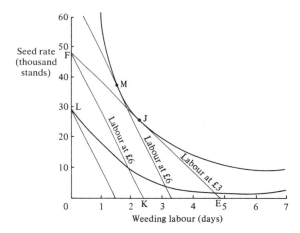

Fig. 2.7 Price change and factor substitution

combination for five bags of maize is now a corner solution at point L. Clearly a rise in the wage provides an incentive to use less labour and more seed; to substitute the cheaper input for the dearer one.

In addition to this 'substitution effect' of the wage rise, the total variable cost of producing ten bags of maize has now risen from £14.70 to £15.60. This rise in the cost of labour will cause a reduction in the economic optimum level of labour input and maize output. This is known as the 'output effect'. Thus when the relative price of labour rises both the substitution and the output effects will lead to a reduction in labour use, but, whereas the substitution effect implies more seed use, the output effect implies less seed use. Whether the two effects in combination will lead to a rise or a fall in seed use is an open question.

An alternative view

Some researchers suggest that input–output relationships are not curved, as we have assumed so far, but are made up of straight lines as shown in Figure 2.8. This is known as 'bent-stick' response for obvious reasons. Because of the practical difficulties of estimating production functions, no one can be sure which theory gives a better description of the real world. In fact Figure 2.8 is similar to Figure 2.2 in that both show diminishing marginal returns. The only significant difference is that now the marginal product

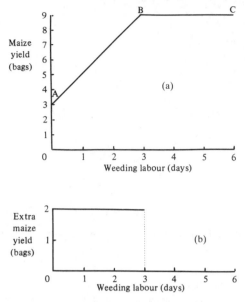

Fig. 2.8 Bent-stick response, (a) Total product
(b) Marginal product

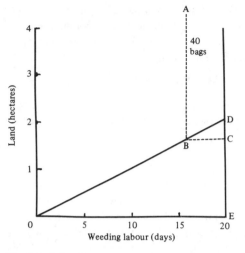

Fig. 2.9 A fixed factor proportions activity

is constant, reflected in the constant slope, until the maximum feasible yield is reached at which point the marginal product suddenly diminishes to zero.

The difference has significant implications for the effect of price changes on the level of input use and yield. Whereas with a smooth response curve the economic–optimum level of weeding labour changes for even small changes in relative prices, this is not the case for bent-stick response. By considering alternative weeding labour cost lines through point B, the reader will see that this represents the economic optimum for wage costs from zero up to two bags of maize per day. However, if the wage should rise above this level, the economic optimum would suddenly shift to point A representing only three bags of maize produced with *no* weeding.

The maximum feasible yield is determined partly by the biological characteristics of the crop variety grown, and partly by the level at which other inputs are fixed. Thus if the seed rate was increased the maximum feasible yield might be raised. However, the law of diminishing returns must still apply; there is an absolute limit to the amount of yield that can be produced from a given plot in a given season, regardless of the amounts of inputs used.

The theory of straight-line or bent-stick production response is usually associated with the assumption that, for a particular productive activity, *no* input substitution is possible. Such a fixed-factor proportions activity may be represented in a two-dimensional diagram like Figure 2.9. This is based on data given in row 1 of Table 2.5. Although this diagram, in common with Figure 2.5, shows the relationship between two inputs, it is quite different in form. First, it may be noted that we are now treating land as one of the variable inputs. This is because, when an activity is expanded, *all* inputs including the land area used must be increased together. Seed use also must be increased but we have omitted this input from the diagram. Second, the ray from the origin OD shows the fixed ratio of labour to land inputs for all levels of production. The line ABC may be viewed as an isoquant but, being right-angled, it simply shows that, since substitution is impossible, an increase in either input *on its own* has no effect on the level of product output.

A productive activity may be expanded until the limited availability of one of the inputs becomes an 'effective constraint'. In Figure 2.9 we assume that the availability of weeding labour is limited to twenty days, represented by the line ED, while land is relatively abundant. Thus labour is the effective constraint and maize production cannot be expanded beyond fifty bags (point D).

Table 2.5. *Alternative activities for maize production*

	Weeding labour per hectare (days)	Yield per hectare (bags)	Weeding labour productivity (bags per day)	Maximum product from twenty days weeding labour (bags)
Activity 1 Hand weeding	10	25	2.5	50
Activity 2 Using herbicides	5	20	4.0	80

New technology

However, there is another possibility for expanding production, namely switching to an alternative method. For instance, the use of herbicides might bring about a substantial reduction in weeding labour. Thus a new activity, 'maize production with herbicides', might be compared with the original as in row 2 of Table 2.5 and in Figure 2.10. Since this new activity requires less weeding labour per hectare of land than the original activity, factor proportions are different and so too is the slope of its ray in the diagram. As a result the area of maize grown can be increased from fifty bags (point D) to eighty bags (point T). Although the yield is assumed to be lower when herbicides are used, the increase in area allows an increase in total maize production.

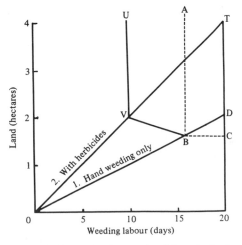

Fig. 2.10 New technology and activity substitution

Now we are admitting that factor substitution is possible after all. By switching from activity 1, hand weeding to activity 2 with herbicides, land and herbicides are substituted for labour. Slightly more land and a lot less labour are used to produce a bag of maize, when herbicides are used. Indeed, if it is assumed that part of the maize crop can be hand weeded while another part is treated with herbicides, combinations of the two activities are possible and an isoquant such as UVBC can be drawn. This is similar to the isoquants of Figure 2.6, although a least-cost combination would occur at a corner representing a single activity, not at a point of tangency.

New technology generally means new methods of production, although it may include product innovations such as new improved crop varieties. Generally speaking, new technology is only worth adopting if it will increase output from a given set of inputs, or reduce costs for a given level of output. If the innovation reduces all costs by the same proportion, its effect is said to be 'neutral'. Thus a new method which increased maize yield per hectare without any change in the quantity of inputs used per hectare would be a neutral innovation. More commonly innovations are 'biased' in that they save some costs more than others. The use of herbicides, in the above example is biased in the direction of saving labour rather than land. It is a labour-saving innovation.

The distinction between 'neutral' and 'biased' technology may also be illustrated for smoothly curved isoquants as in Figure 2.11. In both diagrams T_1T_1 represents the original isoquant while T_2T_2 represents the new technology. Diagram (a) shows the effect of a neutral innovation which simply causes

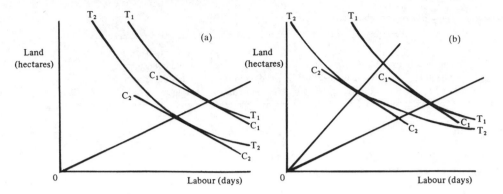

Fig. 2.11 Neutral and labour-saving innovations (relative prices assumed constant: C_2C_2 is parallel to C_1C_1), (a) Neutral innovation, (b) Labour-saving innovation

a parallel shift towards the origin with no change in factor proportions. Diagram (b) shows a labour-saving innovation, which, if prices do not change, results in a bigger reduction in labour use than in the use of land and other inputs. A labour-saving innovation, adopted by many producers, may create unemployment and ultimately cause wages to fall relative to other prices.

There are obviously important implications of this analysis for farm management. A farmer who faces labour shortages or high labour costs, should seek for labour-saving innovations if he wishes to expand his output. If land is the effective constraint he needs land-saving innovations which increase yields and intensity of land use.

Further reading

Adegeye, A. J. & Dittoh J. S. (1982). *Essentials of Agricultural Economics*, Ibadan, Nigeria, Centre for Agricultural and Rural Development (CARD)

Dillon, J. L. (1977). *The Analysis of Response in Crop and Livestock Production*, 2nd edn, Oxford, Pergamon Press

Doll, J. P. & Orazem, F. (1984). *Production Economics: Theory with Application*, 2nd edn, Chichester, John Wiley & Son

Olayide, S. O. & Heady, E. O. (1982). *Introduction to Agricultural Production Economics*, Nigeria, Ibadan University Press

Upton, M. (1976). *Agricultural Production Economics and Resource Use,* Oxford University Press

3

Costs, scale and size

Fixed and variable costs

We have now estimated the economic optimum for a single variable input and the least-cost combination for two variable inputs. This still leaves the problem of finding the economic optimum level of output when there are two or more variable inputs. This problem is best considered in terms of the costs of production, some of which are fixed and some of which are variable. For the example used in Table 2.4 and Figure 2.5 only the costs of seed and weeding labour are assumed to vary as the yield of maize changes. Costs of bush clearing, land preparation and harvesting are all assumed to be fixed in the sense that they do not vary with the yield of maize and must be met anyway. In other words they are unavoidable.

The estimation of these fixed costs would be very difficult in practice since many fixed inputs are neither bought nor sold. However, we shall see that the level of fixed costs has no effect on the economic optimum, except when they are so high that production is not worthwhile. This possibility will be explained later. For the present let us assume that the total fixed cost of cultivating the maize plot is £10.

The total variable cost of producing ten bags of maize (using the least-cost combination of weeding labour and seed) was found to be £11.70 (see Table 2.4 and Figure 2.6). It was also noted, in the previous Chapter, that the least-cost combination for five bags of maize is £4.30. Similarly, the total variable cost can be found for any level of maize output up to about twelve bags, always assuming that the least-cost combination of inputs is used. Total variable costs, estimated in this way for each whole number of bags produced, are given in column 2 of Table 3.1. As might be expected the variable cost increases with

rising output, but the size of the increase gets bigger and bigger. To be more precise, the extra cost of producing one more unit of output, an extra bag of maize in this case, is called the *marginal cost*. Thus the marginal cost increases with rising output, as shown in column 3 of the Table. Note that since fixed costs do not vary with output they do not affect the marginal costs at all. Note also, however, that the reason why marginal costs rise is that some inputs are fixed. These fixed-factor limitations cause diminishing marginal returns as variable inputs are increased, and this is reflected in rising marginal costs. The total cost curves, both fixed and variable are shown in Figure 3.1(a).

Let us now consider the average cost per bag of maize produced. Again we may distinguish the average *fixed cost* from the average *variable* cost. As output is increased, the fixed cost is spread over more bags of maize so the average fixed cost per bag diminishes. However, the rate at which average fixed cost diminishes is rapid to start with but falls as output grows. This effect is shown in column 5 of Table 3.1 and in Figure 3.1(b). The average variable cost, on the other hand, rises as total variable cost rises with increasing output. The overall effect on average cost as output rises is that there is an initial fall, caused by falling average fixed cost, then an increase caused by increasing average variable cost. Thus a typical average cost curve, for both fixed and variable costs together, is U-shaped, as shown in Figure 3.1(b).

The marginal cost curve is also shown in Figure 3.1(b). Note that where marginal cost is less than average cost, the latter is falling; the low marginal cost pulls down the average cost. Conversely, where marginal cost is greater than average cost, then average cost must rise. It follows that marginal and

Table 3.1. *Costs of maize production per plot (£) (fixed cost = £10; price of maize £4 per bag)*

Output bags of maize	Variable cost	Total cost[a]	Marginal cost[b]	Average fixed cost[c]	Average total cost[d]	Gross[e] margin
0	0	10.00				0
			0.80			
1	0.80	10.80		10.00	10.80	3.20
			0.80			
2	1.60	11.60		5.00	5.80	6.40
			0.80			
3	2.40	12.40		3.33	4.13	9.60
			0.90			
4	3.30	13.30		2.50	3.33	12.70
			1.00			
5	4.30	14.30		2.00	2.86	15.70
			1.10			
6	5.40	15.40		1.67	2.56	18.60
			1.20			
7	6.60	16.60		1.43	2.37	21.40
			1.40			
8	8.00	18.00		1.25	2.25	24.00
			1.70			
9	9.70	19.70		1.11	2.19	26.30
			2.00			
10	11.70	21.70		1.00	2.17	29.30
			2.70			
11	14.40	24.40		0.91	2.22	29.60
			5.60			
12	20.00	30.00		0.83	2.50	28.00

[a] Total cost = variable cost + £10 fixed cost.
[b] Marginal cost = difference between successive values of variable cost.
[c] Average fixed cost = £10/number of bags of maize (column 1).
[d] Average total cost = Total cost (column 3)/number of bags of maize (column 1).
[e] Gross margin = Bags of maize × £4 − variable cost.

average costs are equal when average cost is at a minimum. Although not shown in the diagram, average cost may remain constant over a range of outputs. Over this range marginal cost must equal average cost.

Maximizing the gross margin

Profit is the difference between total cost and total revenue, which in our example is the cash earned from sales of maize at £4 per bag. At this stage of the analysis we assume that the farmer's aim is to maximize profit. As in the previous Chapter, the point where profit is maximized is known as the economic optimum. But since fixed cost does not vary with the amount of maize produced, the economic optimum can be found by maximizing the difference between total revenue and total variable cost per hectare. This is known as the gross margin. Gross margins, calculated for each level of maize output, are given in the last column of Table 3.1. The maximum gross margin of £29.60 is obtained when

eleven bags of maize are produced. By subtracting the fixed cost of £10 we can calculate the maximum profit as £19.60. This then is the economic optimum. It is commonly identified as the point where marginal cost is equal to marginal revenue, which is simply the product price. The reader may confirm from the marginal cost column (column 4) of the Table that the nearest values to £4, the price per bag of maize, lie either side of eleven bags output.

This result may be confirmed from the diagrams. Figure 3.2(a) shows the total fixed and variable cost curves with a straight line CB added to represent total revenue. It has a slope of 4 since each bag of maize can be sold for £4. The total gross margin, represented by CA, is at a maximum since the line CB is a tangent to the total cost curve. Lines parallel to CB could be drawn to the left of this line to represent less output, but clearly a smaller gross margin would be earned. Beyond point B, however, the cost curve rises more steeply than the revenue line, so gross margin cannot be increased any further. It may be noted that the slope of the variable (or total) cost

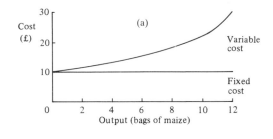

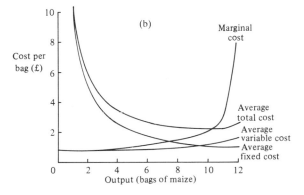

Fig. 3.1 Cost curves, (a) Total cost, (b) Average and marginal cost

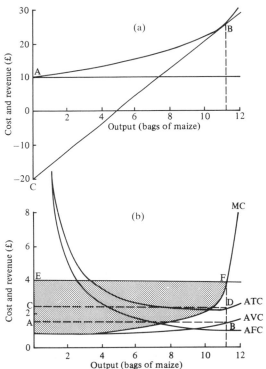

Fig. 3.2 Economic optimum output, (a) Maximum gross margin and profit, (b) Marginal cost equals marginal revenue

curve is the marginal cost, while the slope of the revenue curve CB is the marginal revenue. At a point of tangency the two slopes are equal so we have confirmed that marginal cost equals marginal revenue at the economic optimum. It may also be noted that profit, which equals gross margin minus fixed cost, is represented by CO. This is the maximum profit since the gross margin is maximized.

The marginal and average cost diagram is reproduced in Figure 3.2(b) with a horizontal line EF added to represent the constant price per bag of £4. Marginal cost is equal to marginal revenue at point F at just over eleven bags output. The total gross margin is equal to the shaded area between the marginal cost curve and the revenue line EF. This is maximized at point F. Expanding output beyond this point would add more to costs than it would to revenue so each additional bag produced would make a loss. Once again the condition for an economic optimum is confirmed. Alternatively, in the same diagram, total gross margin is measured by the area ABFE which is simply the average gross margin per

bag of maize (AE or BF) multiplied by the number of bags produced (AB or EF). Similarly, the total fixed cost is measured by the area ABDC, leaving the area CDFE as profit.

The point where marginal cost is equal to the product price is simultaneously an economic optimum for each of the variable inputs. For the present example this means that at 11.3 bags of maize output, where the marginal cost is equal to the price of £4 per bag, the marginal product of weeding labour must equal 0.75 bags of maize as shown in the previous chapter. At the same time the amount of seed used must also be the economic optimum. Figure 3.2(b) may also be used to assess the effect of varying maize price on the quantity a profit maximizing farmer would produce. By comparing points on the marginal cost curve we find that, at a price of £6 per bag, 11.6 bags would be produced whereas, at a price of £2.50, only 10.5 bags would be produced. Thus the marginal cost curve traces out the individual farmer's supply curve relating maize output to price. In this case the curve rises rather steeply implying that price has relatively little impact on quantity supplied; or in other words supply is inelastic. However, this argument only applies so long as average revenue exceeds average total cost; or in other words total revenue exceeds total cost. If, on the contrary, total cost should exceed total revenue, losses would be made. The best or optimum policy might be to abandon production altogether, although this must depend upon what alternative sources of income are available.

Short-run and long-run response

It is helpful in this context to distinguish between short-run and long-run decisions even though the distinction is somewhat arbitrary. We have been considering a short-run decision of how much seed and weeding labour to use and hence what yield to produce on a fixed area of land. Given that the land has already been cleared and many of the fixed costs are already incurred, there is no way of avoiding them this year. It may be worth continuing to grow and harvest the crop even though it will make a loss. So long as the revenue is sufficient to cover the variable cost and a small gross margin is earned it is better than nothing.

In the longer run, however, even by the next season, it should be possible to vary the area of maize grown and, if necessary, substitute some other crop. Thus, other inputs such as the area of land devoted to maize growing, may be varied. With fewer fixed inputs there is less cause for marginal returns to diminish or for marginal costs to rise. If long-run marginal costs do rise, they rise more slowly so the slope is likely to be flatter. In short, it is argued that farmers are likely to be more price responsive the more time they have to adjust. This means that their long-run supply response is more elastic than their short-run response.

Before leaving this subject we should note that the term gross margin has a precise meaning as the difference between revenue and variable costs per hectare of crops or per head of livestock. The term is not applicable if general farm costs, such as the cost of family subsistence, are assumed to vary. Gross margins are used, however, in farm accounting and planning (see Parts III and IV).

Returns to scale

It is arguable that, in some situations, *all* inputs may be varied in the long-run. For instance, where there is a surplus of uncultivated land, the area under cultivation may be expanded to keep pace with population growth. Thus as family size grows so too does the farm size. If all other inputs of seed, manures, tools, livestock and the rest, are increased in proportion, then an 'increase in scale' is said to occur. The effect on output depends upon the 'returns to scale'. Where there are increasing returns, the proportionate growth in output is greater than the proportionate growth of inputs, where there are constant returns the proportionate growth of output is exactly the same as that of inputs, while, where there are decreasing returns to scale, the growth in output is less than that of inputs.

It might well be asked whether an increase in *scale*, that is in strict proportion, ever occurs in practice. After all a growth in family size is likely to result in a change in the land : labour ratio, that is a change in factor proportions. However, if an increase in scale is possible, for instance, if all inputs can be doubled, then we might reasonably expect output to double also; that is for constant returns to apply. In these circumstances marginal and average costs are constant, represented by a horizontal straight line.

It is often assumed that smallholder agriculture, dependent on hand labour, is subject to constant returns to scale. However, there may be advantages in having a large family, which would be reflected in increasing returns to scale and decreasing average costs. For example, where team work is needed in digging a well or clearing dense bush, a large family can provide the necessary team. Also there may be advantages in division of labour since one family member may be away at market while another tends smallstock, another is at work in the fields and yet another is fetching firewood or water. Furthermore, a family with several labour force members, is less likely to be prevented by hazards such as ill health from undertaking the timely planting of seasonal crops (see Hunt, 1984). Yet these advantages, if they exist, are limited especially since small families often share work loads, and provide mutual assistance in times of trouble.

Studies of traditional agriculture in Asia and Latin America have shown an inverse relationship between output per hectare and size of the farm in hectares, after making due allowance for variations in the quality of land (see Berry & Cline 1979). This evidence might be interpreted as demonstrating decreasing returns to scale. Indeed, the authors found that average cost per unit of output was generally lower on the smaller farms, *especially when the price of labour was assumed to be zero.* However, the situation analysed is one of changing factor proportions, rather than differences of scale. Smaller holdings are cultivated more intensively, with higher labour inputs per hectare responsible for the larger output. The average cost per unit of output must depend on the relative values of land and labour. Average costs are only lower on smaller holdings when labour is cheap relative to the value of land. In much of Africa land is less scarce, and therefore less valuable relative to labour than in the Asian and Latin American countries studied. This is related to the fact that, under African communal land tenure, the amount of land cultivated can be adjusted to match up with the family labour supply and food needs. This may result in a shortening of the fallow with consequent lowering of fertility; a case of diminishing marginal returns to increased labour inputs. However, it may be possible to increase farm scale rather than varying the intensity of land use. On balance it seems reasonable to assume constant returns to scale for hand cultivation resulting in constant long-run average costs as shown in Figure 3.3.

However, most observers would refer to all family farming by hand as 'small-scale'. 'Large-scale' production is then taken to mean some form of commercial mechanized system. Here again the difference is not really one of scale, since factor proportions differ between the two systems, large farms using more capital per person employed. Thus capital, mainly in the form of machinery, is substituted for labour. In fact, there is a difference in the technology used, between hand-tools on the one hand and machinery

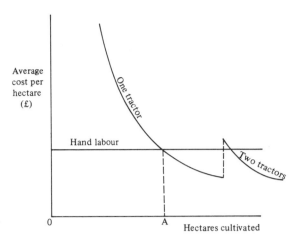

Fig. 3.3 Comparative costs of hand labour and tractor cultivation

on the other. Despite the possible confusion over the precise meaning of 'differences in scale', the terms 'small-scale' and 'large-scale' are so commonly used to describe these different technologies that we will continue to do so.

The technical advantages of 'large-scale' production are generally due to the indivisibility of certain inputs, such as tractors and field machinery, crop and livestock processing plants and irrigation reservoirs. The same applies to expensively trained and skilled manpower. Each item represents a large fixed cost which is spread, or averaged, over the total product output. On a small farm, with a small output, the average fixed cost of such an item must be high, whereas on a larger holding the average fixed cost is lower. This effect of spreading fixed costs on average total cost was illustrated in Table 3.1 and Figures 3.1 and 3.2. Similar average cost curves for tractor cultivation are shown in Figure 3.3.

These curves are only loosely based on field studies (eg see Joy 1960). The situation must differ from one part of Africa to another depending upon the relative availability and costs of labour, credit, machinery, servicing and spare parts and the natural environment. Thus the cost of introducing the plough must be higher in forest areas, where destumping would be involved, than in grassland savannah. None the less, two broad implications of this diagram are generally valid. These are, first, that on very small holdings hand cultivation has the lowest cost, and therefore the highest profit per unit of output and, second, that if the average cost of tractor cultivation falls below that of hand labour it will only do so beyond a certain 'minimum efficient size' of mechanized holding, represented by output OA.

Large-scale production

These advantages of spreading the fixed costs of indivisible inputs such as tractors are the basis of so-called 'technical economies' of large-scale production. They depend on the average cost of the machine technology being less than that of the small-scale alternative. The possible benefits of mechanization include both a 'substitution effect' in saving labour and a 'net contribution effect' in raising output. In practice the substitution effect is limited because weeding and harvesting operations are not easily mechanized. When tractors are used for land preparation, but not for weeding and harvesting, labour requirements are often increased. A net contribution depends upon either increased yields, resulting from deeper tillage and more timely field operations, or extending the area under cultivation. There is little evidence of increased yields being achieved in practice as a result of mechanization. Thus the greatest benefits from tractor mechanization are likely to occur when the cropped area can be increased.

To set against these, often limited, benefits there are relatively high operating costs. Working conditions are generally tough so the costs of wear and tear and damage are high, while lack of spare parts and servicing facilities are frequent problems. Furthermore, there are additional costs of maintaining soil fertility under continual cultivation and possible dangers of increased soil erosion. But large-scale production also suffers from '*diseconomies*' associated with the problems and costs of management and control. Partly for these reasons many large-scale mechanized schemes in tropical Africa have proved to be costly failures, including the East African Groundnut Scheme, State Farms in Ghana and Sierra Leone, and Nigerian Farm Settlements.

In contrast, large-scale commercial farms exist in many parts of Africa and appear to provide their operators with an acceptable margin over costs. However, their apparent success may be associated with 'financial economies' rather than the technical advantages discussed above. First, the large producer often has easier access to markets for his produce. He may well have contractual arrangements to supply a particular buyer at a guaranteed price. Transport costs per unit are generally lower for large shipments of produce. Thus for one reason or another the large-scale producer often receives a higher net price for his produce. Similarly, he is able to buy inputs such as seed and fertilizer in large quantities. Bulk buying may enable him to obtain price discounts and assured deliveries. Probably most important of all the large-scale producer is much more creditworthy than his smaller neighbour. The ownership of capital assets provides security for loans, while the per unit costs of servicing large loans are less than those for small amounts. In short, the large-scale borrower finds it easier to obtain credit, and on more favourable terms, because there are economies of large-scale lending. These financial advantages alone may explain the relative profitability of large-scale farming.

In so far as the costs of marketing and the provision of credit are lower for large-scale producers, society as a whole may benefit from their activities in terms of lower cost production and more efficient use of national resources. But where large producers simply use their political power to change the terms of trade in their favour then society will lose out as resources are diverted away from more efficient users. Other possible social costs of large-scale mechanized agriculture are the gradual degradation and erosion of certain tropical soils under continual monoculture and the displacement of labour when there is no alternative employment. It is argued, on the other hand, that there are important social benefits in transforming agriculture through mechanization; commercial attitudes are engendered, farmers are absorbed into the 'modern' sector and rural–urban migration of school leavers may be discouraged (see Hart, 1980).

It is difficult to generalize regarding the overall impact of large-scale mechanized technology. As remarked already this must vary from one ecological and economic environment to another. Very careful analysis and assessment is needed before policy decisions are made either to promote or discourage this form of production. There are two alternative policy approaches to the promotion of mechanized farming among smallholders whilst possibly deriving some of the financial economies of large-scale trading. These are, on the one hand, the encouragement of producer co-operatives and, on the other, the provision of tractor hire and marketing services by government. The organizational and administrative problems of these institutions will not be discussed here (though see Chapter 9 on management). However, it may be noted that there are major difficulties in imposing a co-operative system on farmers. Indeed some of the large-scale failures mentioned earlier, such as the Nigerian Farm Settlements, were intended to be operated as co-operatives. Most government tractor-hire schemes have operated at a loss, whilst agricultural marketing boards have not always benefited the farmers.

Irrigation schemes, estates and ranches

Brief mention should be made of other large-scale agricultural production systems, namely irrigation schemes, estates and ranches. There are clear technical economies in the construction of large-scale dams for irrigation. A doubling of the height or length of a dam wall increases the volume of water captured by a much bigger factor. Furthermore, because of their more favourable volume to area ratio, evaporation and seepage losses are proportionately smaller from large dams. Similarly, wells, distribution canals and other irrigation works are indivisible items. However, although irrigation is an important means of increasing agricultural production and reducing risk in drier areas, the high costs of construction and the management problems met with on some of the large schemes are causing concern. It would appear that in such cases the diseconomies outweigh the economies of large-scale technology. Governments and funding agencies are becoming more cautious about this form of development. There is increasing interest in the encouragement of small-scale or 'informal' irrigation, which means 'those schemes which are under local responsibility, controlled and operated by the

local people in response to their felt needs' (see Underhill, 1984).

Some perennial crops such as tea, rubber and oil-palms, or crops which occupy the land for several years, like sugar, pineapples and sisal, are produced on estates or plantations, which are large areas of land devoted to a sole crop. The main advantages of this system are the economies of large-scale processing and marketing, often for export. To benefit from these economies, processing factories must be kept working near to capacity and market contracts must be fulfilled. In principle, the most reliable way of achieving these ends whilst minimizing transport costs is to grow the crop on an estate immediately surrounding the factory. All stages from production of the raw material through processing and packaging to marketing of the final product are then vertically integrated into one organization. Tea, coffee and sisal estates exist in East Africa while rubber and oil-palm estates are found in West Africa. Sugar estates have been established in many parts of the continent.

For some of these crops estate production may be the only viable alternative, but not all estates have proved viable. In Nigeria, for instance, state-owned plantations were unprofitable apparently because of a lack of technical data, poor management and high turnover of unskilled labour together with heavy taxation of export crop prices (see Saylor & Eicher, 1970). Efficient small-scale processing methods are available for palm-oil and sugar extraction which may obviate the need for estates. Alternatively, out-grower schemes for smallholders may be combined with public or private provision of central processing and marketing facilities. The Kenya Tea Development Authority is often quoted as a successful example.

In cattle ranching the economies of large-scale operation result from the lower cost per hectare of fencing a large area and the spreading of the costs of providing water points, cattle crushes, dip tanks and disease control measures. However, it is questionable whether the benefits of these measures are sufficient to cover the capital costs which are substantial. Compared with pastoralism there is a large reduction in labour use per hectare and per head of cattle but there is little evidence of any gain in productivity of pasture or livestock, when the value of the pastoralist's milk offtake is included in the comparison (see Cossins & Upton, 1986). Government-owned ranches have performed poorly in most African countries, rarely achieving the planned stocking-rate or predicted reproduction rates. Poor management is blamed for the fact that many such schemes have operated at a loss. Group ranches in Kenya are co-operative enterprises aimed at achieving economies of large-scale operation for owners of relatively small herds. These too have met with administrative and managerial problems. Experiences of this system of livestock production leave doubts as to whether this is a viable alternative to pastoralism.

Further reading

Adams, W. M. & Grove, A. T. eds (1983). *Irrigation in Tropical Africa: Problems and Problem Solving*, Cambridge African Monograph 3, African Studies Centre, Cambridge

Berry, D. A. & Cline, W. R. (1979). *Agrarian Structure and Productivity in Developing Countries*, Baltimore, Johns Hopkins University Press

Cossins, N. J. & Upton, M. (1986). *The Productivity and Potential of the Southern Rangelands of Ethiopia*, Addis Ababa, International Livestock Centre for Africa

Hart, K. (1982). *The Political Economy of West African Agriculture*, Cambridge University Press

Hunt, D. (1984). *The Labour Aspects of Shifting Cultivation in Africa*, Rome, FAO

Joy, J. L. ed. (1960) *Symposium on mechanical cultivation in Uganda*, Kampala, Argus.

Nweke, F. I. (1978). Direct Governmental Production in Agriculture in Ghana, Consequences for Food Production and Consumption 1960–66 and 1967–75, *Food Policy*, **3**(3) 202–208

Saylor, R. G. & Eicher, C. K. (1970). Plantations in Nigeria: Lessons for West African Development. In *Change in Agriculture*, ed. A. H. Bunting, New York, Praeger

Underhill, H. W. (1984). *Small-scale Irrigation in Africa in the Context of Rural Development*, Rome, FAO

4

Choice of products

Mixed cropping

So far we have considered choices relating to a single product; maize. In practice, most African farmers grow a variety of different crops and may also keep some livestock. Thus choices are made between alternative product combinations.

Commonly mixtures of crops are grown on the same plot in the same season, a practice known as 'mixed cropping' or 'intercropping'. Where a crop is planted and harvested, and followed by further crops in the same year, it is described as 'sequential cropping', while if these sequences overlap it is called 'relay cropping'. Mixed, sequential and relay cropping clearly involve choices between combinations of products.

This Chapter is devoted to consideration of how choices are made between alternative products or combinations of them. The example chosen is that of a simple case of mixed-cropping with two crops: maize and cowpeas. We assume that the proportion of the land area devoted to cowpeas can be estimated and that the corresponding yields of each crop are known. The data presented in Table 4.1 are hypothetical, but are not unrealistic. Note that yields are now measured in quintals or decitonnes equal to 100 kg.

These yield estimates are plotted on a graph in Figure 4.1. Such a graph is known as a 'production possibility curve' or 'boundary', since it represents the outer boundary of feasible yield combinations. A farmer who plants too late or who fails to weed his crops may only achieve a yield combination inside the boundary as at point A. Production possibility boundaries may be drawn for any pair of alternative products, crop or livestock or off-farm, and most such curves have certain features in common.

The slope of the curve is a measure of the rate at which one product can replace another, and is known as the 'rate of product transformation' (RPT). In the present example it is the quantity of maize given up or foregone when one more quintal of cowpeas is produced. Over much of the curve the slope is negative because the two crops compete for limited space, light and soil nutrients. The rate of product transformation is therefore defined as *the negative* of the slope of the production possibility boundary. Thus where products are competitive the RPT is positive, as shown by most of the entries in column 4 of Table 4.1.

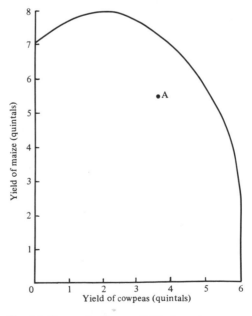

Fig. 4.1 The production possibility boundary

Table 4.1. *Yields under mixed cropping*

Percentage of area devoted to cowpeas	Yield of cowpeas (quintals per ha)	Yield of maize (quintals per ha)	Rate of product transformation[a]
0	0	7.00	
			−0.7
10	1.20	7.80	
			−0.2
20	2.40	8.00	
			0.5
30	3.45	7.50	
			0.9
40	4.30	6.70	
			1.2
50	4.95	5.90	
			1.8
60	5.40	5.10	
			2.3
70	5.70	4.40	
			7.0
80	5.90	3.00	
			16.0
90	6.00	1.40	
			∞
100	6.00	0	

[a] Rate of product transformation calculated as decrease in maize yield divided by increase in cowpea yield.

Where two crops are competitive the expansion of one has a 'cost' in terms of the amount of the alternative foregone. This is an example of an 'opportunity cost', which is a very important concept. Resources of land, labour and home-made capital have opportunity costs even though they may not have a market price. In this example the RPT is really a measure of the opportunity cost of producing cowpeas, in terms of the quantity of maize foregone. Thus the fifth entry in column 4 of the Table, gives the RPT as 1.2. This means that the opportunity cost of producing one extra quintal of cowpeas is 1.2 quintals of maize.

It may be noted that the RPT is also equal to the ratio of the marginal products of the most limiting resource for the two products. Thus the marginal product of the fifth (0.1 hectare) unit of land devoted to cowpeas is $4.95 - 4.30 = 0.65$ quintals. The marginal product of this same unit of land devoted to maize is $6.70 - 5.90 = 0.80$ quintals of maize. Hence the RPT or the opportunity cost of growing one quintal of cowpeas is $0.80/0.65 = 1.2$ quintals of maize. In summary

$$\text{RPT} = dY_m/dY_c = MP_m/MP_c$$

where dY_m = change in maize yield and dY_c = change in cowpea yield; MP_m and MP_c represent marginal products of land (or other limiting resource) in maize and cowpea production respectively.

The first two values of the RPT shown in Table 4.1 are negative which means that the two crops are not competitive over this range. Indeed, it implies that the yield of maize rises with the proportion of cowpeas in the mixture. This may reflect the nitrogen-fixing ability of the cowpea crop. Over this range the crops are 'complementary'. From the bottom of this table it may be noted that up to 10 per cent of the area could be devoted to maize without any loss in cowpea yield. The opportunity cost of introducing a small amount of maize into the mixture is zero. This means that the RPT of cowpeas for maize is infinitely large. For this part of the curve, maize is a 'supplementary enterprise' which does not compete with the cowpea crop.

The production possibility boundary bulges away from the origin; the negative slope is increasingly steep. In other words the RPT increases with increases in the proportion of cowpeas. This is illustrated by the increasing values of the RPT in column 4 of Table 4.1. It is explained in part by complementary and supplementary relationships between products, and in part by the law of diminishing marginal returns. As more cowpea seed is used on a fixed area of land there are diminishing marginal returns so that more and more maize must be given up to achieve a unit increase in cowpea yield. This effect is reinforced since as *less* maize seed is used its marginal product rises. In other words an increasing quantity of maize is foregone for each unit reduction in maize seed.

Constraints and the feasible area

There is increasing awareness among scientists of the ecological advantages of mixed cropping. These are associated with (i) an increase in the utilization of environmental factors such as light, water and nutrients (ii) better control of weeds, pests and diseases and (iii) soil protection (eg see Willey, 1979). The first of these advantages results from the fact that different crops have different water and nutrient requirements and different rooting habits. Thus we may assume that limitations of space, soil moisture and various plant nutrients are constraints on crop growth, but that crops compete at differing rates for

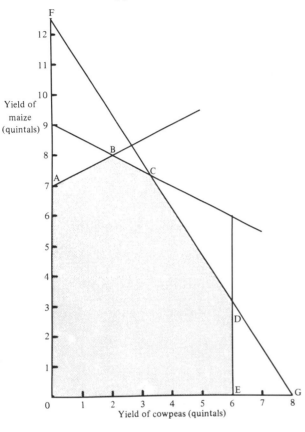

Fig. 4.2 Constraints and the feasible area

Table 4.2. *Constraints on mixed cropping*

	Requirements per quintal of yield		Availability constraint level
	Maize	Cowpea	
Space/light (thousand sq m)	0.8	1.25	10
Available phosphate (units)	3	1.5	27
Available nitrogen (units)	1	−0.5[a]	7
Soil moisture early season (mm)	0[b]	50	300

[a] Negative sign implies that cowpea crop *increases* available nitrogen in the soil.
[b] This constraint only affects cowpeas, since maize is planted later.

them. Then, even if straight-line relationships are assumed, a production possibility boundary rather like Figure 4.1 will result.

To illustrate, it is assumed that there are four main ecological constraints on the growth of maize or cowpeas, which are space, soil moisture, nitrogen and phosphates. Data on availability and requirements of these factors are difficult to estimate, but some hypothetical values are given in Table 4.2. They may be used to construct a production possibility boundary as shown in Figure 4.2.

First consider the space/light constraint. If all the available space were allocated to maize, and there were no other constraints, 10/0.8 = 12.5 quintals could be grown (point F). Alternatively if all the space were allocated to cowpeas 10/1.25 = 8 quintals could be grown (point G). Various combinations of the two crops, subject only to the space constraint,

are represented by the straight line FG. Similarly, the soil phosphate constraint is represented by a line joining 9 quintals of maize (and no cowpeas) to 18 quintals of cowpeas (and no maize), of which BC is a segment.

The case of soil nitrogen is slightly different. If no cowpeas were grown then the maize yield would be limited to 7 quintals (point A). However the introduction of cowpeas *increases* the available nitrogen, so between points A and B each additional quintal of cowpeas produced allows an extra half quintal of maize to be grown. This then illustrates a complementary relationship between the two crops.

Finally we assume that, whereas soil moisture early in the season limits cowpea yield to 6 quintals, maize requires none since it is planted later than the cowpeas. This serves to illustrate the notion of a supplementary enterprise. Between points E and D, maize production can be increased with no corresponding reduction in cowpea yield.

Once again we have a production possibility boundary ABCDE which bulges away from the origin and demonstrates an increasing RPT. The feasible area, representing all feasible combinations of the two products, lies inside or on the boundary. Points outside the boundary are infeasible. Thus choices are restricted by the technical relationships of production.

Maximizing returns

To illustrate the choice of product combinations we return to the smooth boundary of Figure 4.1, although either diagram would serve our purpose. Initially, let us assume that both crops can be sold, maize at £15 per quintal and cowpeas at £20 per quintal, and that the farmer's aim is to maximize his financial returns per hectare. The total revenues calculated at these prices for the crop combinations given in Table 4.1 are presented in Table 4.3. It is clear from these figures that returns are maximized when 50 per cent of the area is devoted to each crop, at which point 4.95 quintals of cowpeas and 5.9 quintals of maize are produced. For this combination the RPT is closest to the inverse price ratio 20/15 which equals 1.33. In fact, the revenue maximizing combination is usually defined as the point where the RPT equals the inverse price ratio.

This is demonstrated in Figure 4.3 which is simply the production possibility boundary with lines of equal revenue, or iso-revenue lines, added. Each of these lines represents all combinations of the two crops which will yield a given total revenue. The maximum feasible revenue is again found at a point

Table 4.3. *Revenues from alternative mixtures*

Percentage of area devoted to cowpeas	Revenue from cowpeas at £20 per quintal (£)	Revenue from maize at £15 per quintal (£)	Total revenue (£)
0	0	105.00	105.00
10	24	117.00	141.00
20	48	120.00	168.00
30	69	112.50	181.50
40	86	100.50	186.50
50	99	88.50	187.50
60	108	76.50	184.50
70	114	66.00	180.00
80	118	45.00	163.00
90	120	21.00	141.00
100	120	0	120.00

of tangency, where the highest possible iso-revenue line just touches the production possibility boundary (point M). Thus the precise optimum product combination is 4.9 quintals of cowpeas and 6 quintals of maize yielding a total revenue of £188.

The slope of a curve at a point is equal to that of its tangent. Since the slope of the iso-revenue line equals the negative of the inverse price ratio, this must equal the RPT at the optimum point. Thus the maximum revenue combination is found where

$$RPT = MP_m/MP_c = R_c/R_m$$

where

$$R_m = \text{maize price}$$

and

$$R_c = \text{cowpea price}$$

Rearranging the equation gives

$$MP_m \cdot R_m = MP_c \cdot R_c$$

which shows that, at the economic optimum, the marginal value product for maize, is equal to that for cowpeas. Thus the general rule for optimum allocation of any factor of production between competing products is to aim for 'equi-marginal returns' in all uses.

These equalities would not hold, if the straight-line production possibility boundary of Figure 4.2 had

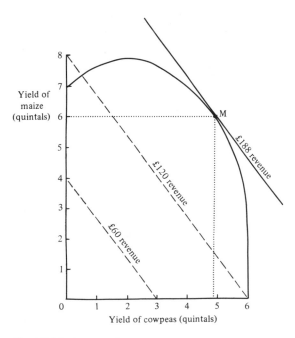

Fig. 4.3 Maximum cash revenue

been used. However, iso-revenue lines could be added and an optimum found. In this case the optimum is at point C representing 3.29 quintals of cowpeas and 7.35 quintals of maize, as the reader may check from the diagram. The important conclusion in this case is that the optimum is a *corner* solution. As we saw in Chapter 2, whereas a tangent solution changes with changing prices, a corner solution is fixed for a range of relative prices.

Subjective preferences

It is convenient, at this point, to drop the restrictive assumption that all products are sold and that farmers aim to maximize their financial gain. We may use the data given in Table 4.1 and plotted in Figure 4.1 to consider how a farmer may choose between maize and cowpeas when these crops are grown for home consumption.

Three alternative theories of decision making will be considered, namely (i) satisficing, (ii) ranking of objectives and (iii) utility maximizing. These all require a little more explanation.

The first is based on the assumption that objectives can be expressed as targets or goals. In the context of our present example, the farmer may aim to produce at least 3 quintals of cowpeas and 5 quintals of maize. These targets may be determined by the basic food needs of the family. They are represented in Figure 4.4 by the lines AC and DE respectively, which now restrict the feasible target area to the shaded region BCE.

However, there are still many feasible combinations of the two crops within this area. So long as the farmer can meet these minimum goals he may not be too concerned about finding a single, best combination or maximizing anything. Any point within the target area is acceptable. Such behaviour is known as 'satisficing'.

If the farm family finds that it is easy to achieve the current set of goals, these goals may be adjusted upwards. But this is a slow adjustment process not driven by the aim of maximizing anything. Although this may be a reasonable explanation of how decisions are taken in practice, it is of limited value for analysis since we cannot identify an optimum or how it might be improved.

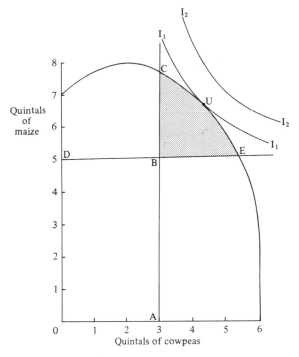

Fig. 4.4 Subjective choice: achieving targets and maximizing utility

The second theory is based on the assumption that farm families can rank their objectives in order of priority. Thus production of sufficient basic, cereal staple for family needs might be the principal objective, while producing other crops for variety of diet may have secondary priority. Earning a cash income, may have even lower priority. The high priority objectives may be treated as goals which must be met. Once a farming system has been found which meets all the high priority goals then choices may be based on lower priority objectives, such as maximizing profits. In our simple example relating to a choice between just two alternatives we may assume that the first priority is to produce the 5 quintals of maize needed to feed the family. In Figure 4.4 this is represented as before by the line DE. The second priority objective might then be to maximize cowpea production, which would be achieved at point E.

This is known as a 'lexigraphic ordering', meaning the approach used by dictionary makers. Thus, in a dictionary, words are arranged first in order of the first letter, then in order of the second letter, and so on.

Utility maximization

For the third theory it is assumed that the farmer's total welfare or satisfaction can be expressed as a quantity of 'utility'. His total utility is a function of the quantities of goods consumed or other needs satisfied, just as we earlier assumed that farm production is a function of the quantities of inputs used. Given these assumptions we may consider contours of his utility function which are known as 'indifference curves'. They have many features in common with isoquants, discussed earlier, but, whereas these represent different combinations of inputs which yield a constant physical output, an indifference curve represents different combinations of a pair of goods or services which yield a constant level of utility. The curves I_1I_1 and I_2I_2 in Figure 4.4 are indifference curves.

The name derives from the conclusion that, since all points on a curve yield equal utility, the individual concerned must be indifferent between the combinations of goods they represent. Just like isoquants, they generally slope downwards (have a negative slope) since more of one good is needed to compensate for less of the other.

They bulge towards the origin, which means the rate of substitution diminishes along the curve. The reason may be illustrated from our example, where cowpeas are substituted for maize. If our farmer were free to exchange maize for cowpeas we might expect that as the quantity of cowpeas increased he would get less and less utility from each additional kg; cowpeas would yield diminishing marginal utility. At the same time as he gave up more and more maize its marginal utility would increase. Thus the amount of maize he is willing to forego in exchange for an additional kg of cowpeas, gets smaller and smaller.

There is a different indifference curve for every different level of utility an individual might attain. The further away from the origin a particular indifference curve lies, the higher is the level of utility it represents; I_2I_2 represents a higher level of utility than I_1I_1. Hence utility is maximized where the highest attainable utility curve is tangent to the pro-

duction possibility boundary as at point U in Figure 4.4.

This analysis suffers from the defect that the concept of a utility function and its corrésponding indifference curve really relates to a single individual. Since individuals differ in their tastes and preferences, so too do their utility functions. In family farming decisions must be arrived at jointly, through some combination of individual preferences. The use of a single set of indifference curves to represent this joint decision-making is clearly a simplification of reality. None the less, all three choice theories discussed are of some value in helping our understanding of the decision process. They are all three used in the Chapters which follow.

Food versus cash crops

A similar analysis may illuminate the choice between food and cash crops, and the effect of rising prices for the latter. We now assume that, although not necessarily committed to mixed cropping, our farmer is endowed with limited resources. Hence he must choose between competing uses of these resources, maize growing for food and cotton growing for cash. The production possibility boundary showing alternative feasible combinations of food production and cash earnings (at the initial price level) is shown by the curve AB in Figure 4.5. The indifference curves now represent the farmer's subjective trade-off between food and cash. His optimum choice is at point C.

If the price of the cash crop rises, the returns from cash crop production must also rise and the opportunity cost of food production increases. For instance suppose the price of cotton rises by 50 per cent, then the feasible combinations of food and cash production are now bounded by the line AD where OD is 50 per cent greater than OB. The farmer's optimum now lies on a higher indifference curve I_2I_2 at point E.

The total effect of the price rise on the pattern of production and consumption is made up of two parts, a substitution effect and an income effect. The former occurs because cash production has become more rewarding in comparison with food production, which provides an incentive to substitute cash crops

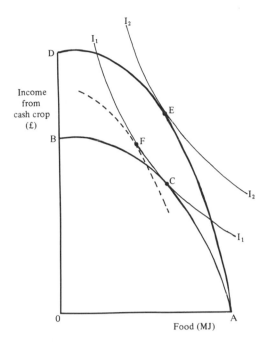

Fig. 4.5 The food crop–cash crop choice

substitution and the income effects will tend to increase sales (in Figure 4.5, F represents a higher level of cash crop sales than C, while E represents a higher level than F). This analysis does not make clear whether the physical output of the cash crop will rise, the increased value may simply be due to the rise in value per unit. However, the majority of empirical studies suggest that tropical smallholders do increase supply in response to an increase in crop price, though this may be largely due to substitution of one cash crop for another. (See Askari & Cummings, 1976.)

In practice, there are choices to be made not simply between cash production and food production but between a whole range of consumption goods and a large array of alternative production processes. Similar principles of individual choice are supposed to apply. However, it should be remembered that a decision to produce basic food needs on the farm implies adoption of a range of different crops and storage activities in order to provide a balanced diet throughout the year.

for food crops. The latter occurs since the price rise causes farm incomes to rise and the farmer may respond by consuming more food than before at the expense of having less to sell. The total effect of a rise in cash crop price is the sum of these two effects but depends on their relative strengths, since one is positive and the other negative.

These two effects are separated in Figure 4.5, by constructing the broken line, through point F, parallel to AD. This represents the change in relative values of cash and food crops, without any corresponding increase in the farmer's total satisfaction; which would give rise to a new optimum product combination at F. Now the substitution effect would be a move from C to F and the income effect a move from F to E. Since the income and substitution effects may act in opposing directions, the total effect on food production is unpredictable. In this hypothetical example, the rise in cash crop price has no impact on food production (point E is vertically above point C), since the income effect just cancels out the substitution effect. The impact of the price rise on cash sales is more likely to be positive since both the

Specialization versus diversification

In the last Chapter we considered the possible economies of large-scale production. Where such economies exist for a particular productive activity, there are advantages in specialization and devoting all available resources to that activity. In this Chapter, however we have reviewed the case for diversification into more than one productive activity. In summary there are several potential advantages.

First, given that some inputs are fixed, marginal returns to variable inputs are likely to diminish as more and more are devoted to a single product. Higher marginal returns and hence more total product are obtained by devoting some of the inputs to an alternative crop or livestock activity.

Second, complementary and supplementary relationships between alternative products mean that the combined output of both from a given set of resources is greater than that of either one on its own.

Third, if in fact there are several constraints as shown in Figure 4.2, then, it is unlikely that a single productive activity, will yield the maximum return. It can be proved (see Chapter 16) that the number of activities in an optimal plan must equal the number of effective constraints, which are actually limiting production. Generally speaking, total production can be increased by exploiting available resources up to their limits, which means making the constraints effective. This then will involve a combination of productive activities.

Fourth, where subjective choice is involved, as in choosing what crops to grow for home consumption, a variety of products is likely to be preferred over a single one.

Finally, risk may be reduced by diversification but this will be discussed in the next Chapter.

Further reading

Askari, H. & Cummings, J. (1976). *Agricultural Supply Response: A Survey of the Econometric Evidence*, New York, Praeger

Lightfoot, C. W. F. (1980). *An Initial Report on the Evaluation of Intercropping*, Development Plan Number ARO8, Botswana, Ministry of Agriculture

Norman, D. W. (1974). The rationalisation of a crop mixture strategy adopted by farmers under indigenous conditions: the example of Northern Nigeria, *Journal of Development Studies*, 11 pp. 3–21

Willey, R. W. (1979). Intercropping: its importance and research needs, *Field Crop Abstracts*, 32, Nos. 1 and 2, pp. 1–85

Also see suggestions for further reading at the end of Chapter 2.

5

Risk avoidance

Uncertainty in agriculture

A farmer, when he embarks on any productive activity, is uncertain what the actual outcome will be. Uncertainty has three main causes (i) environmental variations causing production and yield uncertainty, (ii) price variation causing market uncertainty and (iii) lack of information. All of these are significant in African agriculture, where unreliable rains and pest and disease outbreaks cause wide variation in resource availability and in crop and livestock yields. Human diseases are frequent, unpredictable and costly to treat. Ill health or injury of a family member at a critical period may cause serious loss of production and income. Generally there are wide seasonal and unpredictable fluctuations in market prices, while information on alternative technologies or the market situation outside the immediate locality is often lacking. Hence the farmer cannot plan with certainty; his decisions are subject to risk.

Risk is a measure of the effect of uncertainty on the decision-maker. There are differences of opinion as to how risk should be measured. Some argue that it is variation or instability of income, while others claim that it is the possibility of disaster or ruin. Both these alternatives will be explored. In any case there is fairly general agreement that most people, including farmers, are risk averse. This means that they are willing to forego some income or face extra costs in order to avoid risk. They are cautious in their decision-making.

These ideas are best illustrated by a simple example comparing the hypothetical returns from two alternative crops, maize and sorghum, with variable yields. The so-called 'pay-off matrix' is given in Table 5.1. Since we are only concerned with yield uncertainty here, the pay-offs are in quintals of grain rather than in cash terms, a quintal of sorghum being valued equally with a quintal of maize. The pay-off from a particular activity is supposed to depend upon the 'state of nature' which obtains during the subsequent cropping season. We have further simplified the analysis by assuming only three possible states of nature; wet years, normal years and dry years. In practice, there is an enormous range of possible states of nature. The farmer does not know which of these will occur. To start with we assume he does not know the likelihood of their occurring.

The figures have been chosen to show that sorghum yields are less variable than maize yields. Also whereas maize yields are higher in wet years, sorghum yields are higher in dry years. If a choice had to be made between the two crops, a cautious farmer would clearly choose to grow sorghum as the less risky. He is then assured of a yield of at least 40 quintals of grain even under the worst conditions. This is known as the MAXIMIN choice since it maximizes the minimum or worst possible outcome.

This is really a *very* cautious policy, since it assumes that 'nature' will always do her worst. By growing sorghum the farmer foregoes 30 quintals of grain in wet years and 10 quintals in normal years, while he

Table 5.1. *A pay-off matrix (yields in quintals of grain per ha)*

	States of nature		
	Wet years	Normal years	Dry years
Maize	70	60	30
Sorgum	40	50	50

only gains 20 quintals in dry years. There is, in effect, some trade-off between net gains and risk avoidance. The choice would be clear cut, however, if one crop, say maize, yielded more than the other under *all* states of nature, in which case maize production would 'dominate' sorghum. In this example neither activity dominates.

Probability and expectations

In practice, the farmer's choice is likely to be influenced by the probabilities of occurrence of the different states of nature. Clearly, if dry years occur only one year in ten, that is their probability of occurrence is only 0.1, sorghum is less likely to be chosen than if dry years occur more frequently. It is reasonable to assume that farmers can judge the probabilities of wet and dry years or different states of nature on the basis of their past experience and that these judgements influence their choices. Incidentally, the distinction which used to be made between 'risk', where probabilities are known, and 'uncertainty' where probabilities are not known, is no longer accepted since it is argued that decision-makers have to make judgements of the relevant probabilities in every risky situation.

Returning to the example, let us assume that dry years are judged to occur three years in ten on average, while wet years occur two years in ten. Since there are only three states of nature and their probabilities must sum to one, the probabilities of wet, dry and normal years are 0.2, 0.3 and 0.5 respectively. The expected yields for the two crops can now be calculated using the formula

$$E(Y) = \sum Y_iP_i = Y_1P_1 + Y_2P_2 + \cdots + Y_nP_n \quad (1)$$

where the amounts Y_1, Y_2, . . . or Y_n are the yields under the various states of nature and $P_1, P_2, . . .$ and P_n are the respective probabilities and the Σ sign means the sum of n similar terms as shown. The expected yield for maize is therefore

$$70 \times 0.2 + 60 \times 0.5 + 30 \times 0.3 = 53 \text{ quintals}$$

and for sorghum

$$40 \times 0.2 + 50 \times 0.5 + 50 \times 0.3 = 48 \text{ quintals}.$$

If the farmer were risk indifferent, rather than risk averse, he would choose maize production since it yields the highest expected return. Indeed, this is arguably the only rational choice since the expected value is simply an average. In the long run, better than average years will cancel the effect of worse than average years so, provided the family survives, on average, returns will be maximized in this way. However, if worse than average years could prove disastrous, it is perfectly rational to avoid risk.

Given estimates of the probabilities of different outcomes, variation can be quantified by either the variance, which is the square of the 'standard deviation'

$$V(Y) = \sum (Y_i - E(Y))^2P_i = E(Y^2) - (E(Y))^2 \quad (2)$$

(note the two alternative methods of calculation) or the mean *absolute* deviation

$$MAD(Y) = \sum |Y_i - E(Y)| \cdot P_i \quad (3)$$

The variance and mean absolute deviation of yield per hectare of maize (Y_1) are therefore

$$V(Y_1) = 17^2 \times 0.2 + 7^2 \times 0.5 + 23^2 \times 0.3$$
$$= 241$$
$$MAD(Y_1) = 17 \times 0.2 + 7 \times 0.5 + 23 \times 0.3$$
$$= 13.8$$

while for sorghum yields (Y_2) they are;

$$V(Y_2) = 8^2 \times 0.2 + 2^2 \times 0.5 + 2^2 \times 0.3 = 16$$
$$MAD(Y_2) = 8 \times 0.2 + 2 \times 0.5 + 2 \times 0.3 = 3.2$$

as the reader may care to check.
Both measures lead to the same conclusion, namely that sorghum has less variable yields than maize. Of course, this is obvious from inspection of Table 5.1 without making these calculations. However, when analysing more complicated, real-world situations it is useful to have some measures of variation. Generally, both measures described here give the same ranking of alternatives.

Minimizing variation of returns cannot be the farmer's sole objective. After all he could reduce variation to zero by producing nothing at all. We must therefore assume he has a second objective of maximizing expected or average return. Clearly, he has a difficult choice. Expected return is maximized by growing maize, while the yields of sorghum are less variable.

Diversification and risk

An obvious possibility, not yet considered, is to grow both crops in a mixture. For simplicity we now assume a constant rate of product transformation between the two crops. This means that for a given state of nature, the total grain yield $Y(T)$ from a combination of K_1 hectares of maize and K_2 hectares of sorghum is given by

$$Y(T) = K_1 Y_1 + K_2 Y_2 \qquad (4)$$

This equation is readily extended to cover cases where there are more than two alternative activities. In our example we assume only one hectare of land is available so $K_1 + K_2 = 1$.

Using equation (4) and the data given in Table 5.1, yields can be calculated for various combinations of the two crops under the alternative states of nature. Results are presented in Table 5.2. It is apparent that the yields from mixtures of the two crops are less variable than yields of either crop on its own. Further analysis of the effect of diversification on risk may be pursued in two ways (i) in terms of worst possible outcomes or (ii) in terms of variation of outcomes.

Perusal of Table 5.2 will make it clear that for sorghum-based mixtures with less than 20 per cent maize, the worst outcomes occur in wet years. For mixtures with more than 20 per cent maize the worst outcomes occur in dry years. For a mixture of exactly 20 per cent maize and 80 per cent sorghum the lower yield of 46 quintals is obtained in wet or dry years. Furthermore, this is the maximum or best of the

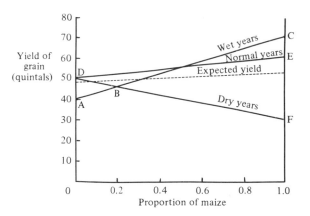

Fig. 5.1 The maximin mixed strategy

worst outcomes. It therefore represents the 'MAXI-MIN' mixed strategy.

This result is illustrated in Figure 5.1 where the horizontal axis represents the proportion of maize in the combination, while yields are measured on the vertical axis. The lines AC, DE and DF show how total yields vary with an increasing proportion of maize in wet, normal and dry years respectively. The minimum outcomes for all combinations of the two crops are given by the line ABF. Clearly point B represents the maximum level of these worst yields, that is the maximin point.

As already noted, a maximin strategy has an opportunity cost in terms of reduced expected, or average, returns. The total expected return $E(T)$ from a mixture of activities can be calculated from the expected yields of the component activities using equation (4). In the present two-activity example, maize has the higher expected yield, so total expected return increases directly with the proportion of maize in the mixture. The trade-off with risk may be expressed by plotting minimum return against total expected return as in Figure 5.2 which is derived directly from Figure 5.1. Between points B and F, expected return can only be increased by accepting a lower minimum. A farmer who is very risk averse might choose point B, while one who is risk indifferent, may choose to maximize total expected yield at point F. Most farmers would probably choose some intermediate point on the line BF.

The alternative theories of choice discussed in the

Table 5.2. *The effects of diversification (yields in quintals of grain per hectare)*

| Percentage of maize in mixture | States of nature | | | |
	Wet years	Normal years	Dry years	Expected value[a]
0	40	50	50	48
20	46	52	46	49
40	52	54	42	50
60	58	56	38	51
80	64	58	34	52
100	70	60	30	53

[a] Based on $p(\text{wet}) = 0.2$; $p(\text{normal}) = 0.5$; $p(\text{dry}) = 0.3$.

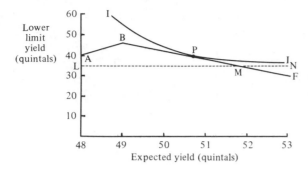

Fig. 5.2 The expectations–lower limit trade-off

last Chapter may be applied to risky choices. One approach is to assume that the farmer has a target acceptable level of return, which may be the minimum necessary for survival, that he aims to achieve in all years, even the most adverse, drought years. To illustrate let us assume the target minimum yield is 35 quintals of grain, which is represented by the horizontal line LMN in Figure 5.2. This now restricts choice to crop combinations with not more than 75 per cent maize. A 'satisficer' might be content with any combination which will achieve this target.

A closely related alternative is to assume that the farmer is a lexicographic optimizer, who aims at meeting the target minimum survival level as first priority and maximizes expected returns as second priority. This is known as a 'safety-first strategy' for obvious reasons. In our example this would lead to a choice of point M representing 25 per cent sorghum and 75 per cent maize, as the optimum. The total expected yield for this combination is $0.75 \times 53 + 0.25 \times 48 = 51.75$ quintals.

Alternatively, we could assume that the minimum income is a variable in the farmer's utility function. His indifference curves, such as II in the diagram then determine his optimum, in this case at the point of tangency P. The farmer's attitude to risk is reflected in the slope of the indifference curves; the more risk averse he is, the flatter are his indifference curves and the nearer to the maximin combination is his optimum.

It should be noted that, in more realistic analyses with many more possible states of nature, such concern with the worst possible outcome may be

inappropriate, since the probability of this outcome may be very small indeed. A more appropriate criterion for the 'lower limit' might be that there is a relatively small probability of falling below it. Indeed, the objective could be formulated as minimizing the probability of falling below the disaster level or maximizing the probability of survival. Yet another alternative is to assume the farmer can identify a 'focus-loss' which is the most unfavourable outcome he will consider. This is not necessarily the worst possible outcome since results which are very unlikely may be ignored.

Variance minimization

We can broaden our understanding of the effect of diversification and risk avoidance, if we now assume that the farmer's aim is to minimize variance, rather than to avoid disasters.

Diversification does not always reduce variance. It all depends upon the relationship between the yields of the different activities, which is measured by the covariance or the correlation coefficient. The covariance between variables Y_1 and Y_2, $COV(Y_1Y_2)$ is given by:

$$\begin{aligned} COV(Y_1Y_2) \\ = \sum (Y_{1i} - E(Y_1))(Y_{2i} - E(Y_2))P_i \\ = E(Y_1Y_2) - E(Y_1)E(Y_2) \end{aligned} \quad (5)$$

The correlation coefficient, denoted by r, is then

$$r = COV(Y_1Y_2)/\sqrt{(V(Y_1)V(Y_2))} \quad (6)$$

which has a maximum absolute value of one and a minimum of zero. For the data given in Table 5.1 the covariance is -34 and the correlation coefficient is -0.55.

The negative correlation between crop yields, obtained in this case, means that they move in opposite directions: when the yield of one is high the yield of the other is low. In these circumstances variance of total yield must be reduced by diversification. The effect on variance of total yield $V(T)$ may be calculated as follows

$$V(T) = K_1^2V(Y_1) + K_2^2V(Y_2) + 2K_1K_2COV(Y_1Y_2) \quad (7)$$

where K_1, and K_2 are the areas of each of the two

crops as before. This form of equation is known as a 'quadratic' function and is readily extended to allow for more different crops, in the mixture.

We can now see from this equation that where the covariance is negative so too is the last term so that the combined variance V(T) is less than the sum of the individual variances $K_1^2 V(Y_1) + K_2^2 V(Y_2)$. But even where the yields are independent, so the covariance is zero, the combined variance will be reduced by diversifying as may be shown if we assume $V(Y_1) = V(Y_2) = V$ and $K_1 = K_2 = \frac{1}{2}$. Then substituting in equation (7) gives:

$$V(T) = (\tfrac{1}{2})^2 V + (\tfrac{1}{2})^2 V = V/4 + V/4 = V/2,$$

so that the variance of the combination is only half that of each individual crop.

Variance *may* also be reduced in this way even when the correlation is positive. However, it is likely to be *increased* by combining crops with strongly positive correlation between yields, as often occurs in practice; all crops fail together in a drought, while all yields are high in favourable years. Whether diversification will reduce variance of total yield, actually depends upon whether the covariance is smaller than the yield variance for each of the crops on its own.

Maximizing expected utility

Let us now return to our original example of Table 5.1 and consider the trade-off between variance and expected values. These have been calculated for various combinations of maize and sorghum and are presented in Table 5.3. Since we assume that resource constraints limit the farmer to growing only one hectare of the two crops (so that $K_1 + K_2 = 1$) the data recorded in the Table, other than for sorghum alone, represent points on the expectations–variance (E–V) boundary, which is plotted in Figure 5.3. This curve is a boundary or frontier since points inside and to the left of it, such as point A representing a pure stand of sorghum, are feasible but inefficient. A larger expected return with a smaller variance may be obtained by choosing a combination on the frontier. From the Figure we see that variance is minimized by combining 0.154 of a hectare of maize with 0.846 of a hectare of sorghum (Point B). This point does *not* coincide with the maximin combination, although it

Table 5.3. *Diversification, variance and mean absolute deviation*

Percentage of maize in mixture	Expected value	Variance	Mean absolute deviation
0	48	16	3.2
20	49	9	3.0
40	50	28	4.8
60	51	73	7.8
80	52	144	10.8
100	53	241	13.8

still represents a low expected return and a very cautious choice. Mean absolute deviation measures are also given in Table 5.3 and plotted in Figure 5.3. The minimum mean absolute deviation combination is very close to that of minimum variance, and, as already mentioned, the two measures usually lead to similar results. However, there is one important difference. Whereas the variance of any combination of activities can be calculated directly from a knowledge of the variances and covariances using equation (7), calculation of the mean absolute deviation, or indeed the worst possible outcome, involves estimation of the return under each state of nature using equation (4) before combining them in a single measure of risk. Thus the variance of returns from a combination of activities is easier to estimate than other measures of variation. However, 'linear programming' can be used to find the *minimum* mean absolute deviation while finding the *minimum* variance, requires more complicated 'quadratic programming' (see Chapter 16).

Choices between expected return and variance are usually viewed in terms of a utility trade-off. Both the utility of extra expected returns and the disutility of increased variance are supposed to be included in the farmer's utility function. Thus indifference curves can be drawn, such as QPI in Figure 5.4, to show the subjective trade-off between expected value and variance. The optimum then occurs at the point of tangency, P. This point represents an expected yield of 50.5 quintals of grain, which is below the maximum feasible expected yield of 53 quintals. The difference of 2.5 quintals is the amount of yield the farmer is

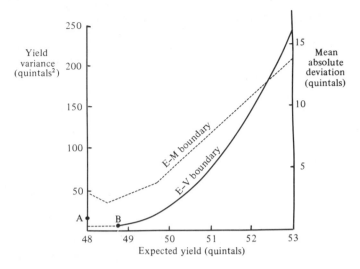

Fig. 5.3 The expectations–variance (E–V) boundary and the expectations–mean absolute deviation (E–M) boundary

willing to forego, on average, to reduce risk. Indeed, he would give up even more in order to avoid risk entirely. Point Q is on the same indifference curve as point P, so it represents the same level of utility, but at point Q the variance is zero, so the outcome is certain. Thus point Q may be described as the 'certainty equivalent' of point P for this farmer. The amount QR (70 kg of grain) is what the farmer would be willing to pay, say for crop insurance, in order to eliminate risk entirely. This is the cost of risk for him, or his risk premium if expressed as a fraction of his expected return.

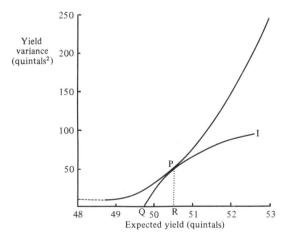

Fig. 5.4 Expected utility maximization

Much research has been done in recent years, involving the estimation of utility functions of this type for individual farmers, in order to measure their attitudes to risk. The estimation procedure is based on farmer interviews in which the respondent is asked to choose between alternative hypothetical gambles, with different sets of prizes and/or different probabilities. (Further details are given in Anderson, Dillon & Hardaker 1977.)

Choices under risk

From this discussion, it appears that there are several alternative theories, or models, of decision-making under risk. Some are based on the concept of a lower limit or disaster level income, while others are concerned with income variation. There is no general agreement over which model is the most useful.

Some have argued that none of the models is very convincing as a description of the way decisions are made in practice. Given the complexity of the real world. decision-makers may not be aware of all the possible states of nature and/or possible strategies. It is even less likely that they can estimate probabilities of all states of nature and rank alternatives so as to find an optimum. Thus is might be concluded that most people are cautious sub-optimizers or satisficers, guided by fairly crude 'rules of thumb'.

The counter-argument is that, although decisions are not consciously made according to these theoretical models, the rules of thumb used in practice lead to similar results. Decisions are made *as if* an optimum is sought.

Whatever the case, these theories have confirmed, first, that risk aversion has a cost in terms of income foregone on average, second, that some activities are more risky than others and may therefore be avoided, and, third, that diversification reduces risk provided that the component activities are not strongly positively correlated. Thus risk aversion may explain why farmers are reluctant to adopt new crop varieties with high but variable yields, why they prefer to grow subsistence crops even when food may be purchased more cheaply, why they practise mixed cropping and why they pursue many on- and off-farm activities. However, there may be other good reasons for adopting these policies. Expected returns may be increased as was shown for the case of mixed cropping in the last Chapter. We should be cautious in imputing farmers' motives from their observed behaviour.

A study of 1500 smallholder farmers in Kenya (Wolgin, 1975) provided rather stronger evidence of risk averse behaviour. First, it was found that the marginal value products for most inputs were higher than their unit costs. This implies that farmers used less than the economic optimum level of input (see Chapter 2) and may be explained by farmers' willingness to forego income in exchange for a reduction in risk. Stronger evidence that this was the main reason is provided by the fact that the ranking of crops by marginal value product correlated closely with their marginal contributions to risk. In other words, the riskier the crop, the higher was the marginal value product, which implies that inputs were further below the 'economic optimum'.

In so far as farmers do take precautions against risk, they will differ in the extent of their risk aversion. This may cause or exacerbate income and wealth disparities between households. The cautious risk avoider will pay more, or forego more income on average, than will the expected value maximizer. Given that the wealthy farmer, with plentiful resources, can afford to take risks and innovate while his poorer neighbour cannot, there is a natural tendency for the rich to get richer while the poor may stagnate. Financial assistance to the poor may enable them to become less risk averse and more innovative.

Sequential risks

In practice, the final outcome of a risky decision generally depends upon a sequence of events. Thus the crop yield obtained depends upon the soil moisture status at several stages of the growth season, for instance (i) at planting and germination and (ii) at grain fill. Furthermore, the moisture availability at grain fill may depend upon the moisture available at planting. The probability of a given moisture level at grain fill is then *conditional* upon the moisture level at planting. In such circumstances, a tree diagram, such as Figure 5.5, is useful for estimating the probability distribution of final outcomes.

In this diagram, the probabilities of different moisture levels are written alongside the relevant branches, those for the later period being conditional probabilities (the conditional probability of event B given that event A has occurred is written as $P(B/A)$ while the probability of both A and B occurring is written as $P(AB)$). The probabilities of the final outcomes are calculated by multiplying the probabilities together along each branch as shown in Figure 5.5(a) (that is $P(AB) = P(A) \cdot P(B/A)$). Alternatively an expected value can be calculated for each node, or junction, as shown in Figure 5.5 (b).

This extension of the theory makes little difference to the analysis, except that each state of nature is now seen to represent a series of events or outcomes. However, care must be taken in using simple averages to estimate expected values, where events are not independent. For instance it would be wrong to calculate expected revenue per hectare as the product of average yield and average price, if the price received is dependent on the yield obtained (see Upton & Casey 1974).

Actually, each crop harvest is not a single isolated event. Rather, it is one of a sequence of harvests. A farmer's ability to survive a poor season may well depend upon the size of the previous harvest. To illustrate, let us consider a very simple situation, with just two possible outcomes for each cropping season; a good crop with probability 0.8 and crop failure with probability 0.2. Part of the tree diagram is shown in

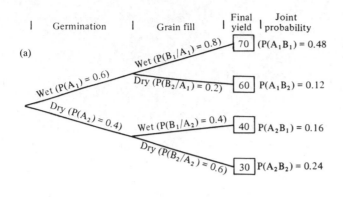

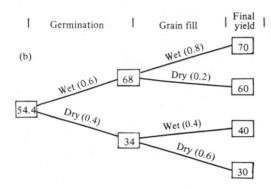

Fig. 5.5 Tree diagrams, (a) Joint probabilities,
(b) Expected values

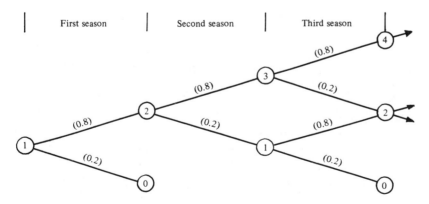

Fig. 5.6 Sequential risks of disaster

Figure 5.6. We assume that the probabilities are *not* conditional on the outcome of the previous season, although they might well be in practice. The circled numbers at the nodes of the figure represent the level of reserves, or the number of seasons the family can survive, not the expected values as in Figure 5.5(b). Obviously the more reserves are carried, the greater are the chances of survival. With reserves for only one season the probability of disaster is 0.2/ 0.8 = 0.25 whereas with reserves to cover two seasons, the probability of disaster is reduced to 0.25 × 0.25 = 0.065. Most cultivators and pastoralists try to limit risk by accumulating reserves of produce or money to survive adverse seasons. (Those interested in pursuing this kind of analysis, referred to as 'gambler's ruin' should read Feller, W. (1950). *An introduction to probability theory and its applications*, New York, Wiley.)

Multistage decision-making

In practice the farmer does *not* make all his decisions at one point in time, then wait for nature to take its course. Rather, decision-making is a sequential process. The final crop yield obtained is the outcome of a series of decisions of when and how much seed to plant, when to weed and how much labour to use, when to harvest and so on. Thus the farmer constantly adapts and adjusts to a changing environment. Indeed, many African farmers purposely adopt a piecemeal approach to their decisions in order to maintain flexibility. Typically, only a part of the cropped area is planted at the first rains, decisions about cropping of the remainder being delayed until more is known about the pattern of the weather.

Such multistage decision problems may be represented by decision trees, a simple example of which is given in Figure 5.7. It shows an alternating sequence of actions and outcomes, starting from the left and moving across to the right. Actions stem from decision modes represented by black squares, while outcomes stem from chance modes which are unmarked. As before, probabilities are shown (in brackets) alongside outcome branches. In this example, the initial decision is whether to plant at the outset of the rains. The second decision whether to interplant, extend the cropped area, or plant for the

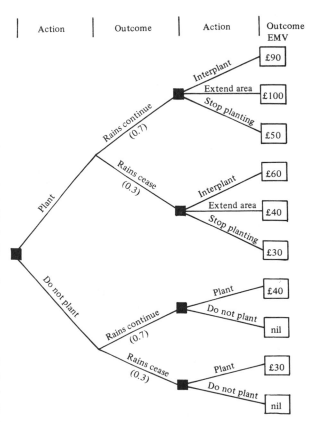

Fig. 5.7 A multistage decision problem

first time is taken a month later, in the light of whether the rains have continued. For simplicity, the decision tree finishes at this point, although in practice further decisions would arise before the final yields are obtained. A three-stage decision tree of this type was used in a World Bank study in Northern Nigeria to show how a multiplicity of different crop sequences and mixtures can result (Balcet, 1982). In our example hypothetical expected money values (EMVs) for each branch are shown (boxed in) on the right-hand side of the diagram.

The optimal sequence of decisions is determined by working backwards from the right-hand side using a process of 'averaging out and folding back'. This means that at each decision mode, the branch with the highest EMV is chosen, while at each chance node the EMV is averaged out. Figure 5.8 which is derived from Figure 5.7 illustrates the process. From

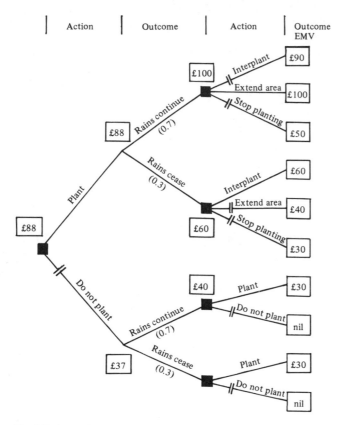

Fig. 5.8 Averaging out and folding back

the upper branches on the right, it is clear that the EMV from extending the planted area is greater than that from interplanting or stopping. Hence the latter options are rejected and the EMV at the decision node is the same as that for extending the area. Now the EMVs for the two possible outcomes can be averaged out at each chance mode. Since the EMV at the planting mode is the greater, this is the option chosen. The optimal sequence of decisions in this case is to plant at the first rains and then to interplant, this sequence giving an EMV of £88.

This method of analysis of multistage decisions is known as 'Dynamic Programming'. It depends upon the principle that the optimal policy at any given stage depends only on the present state of the process, and not on how that state was achieved. As a result decisions can be considered one at a time, working backwards from the final decision. The method can

be used to analyse very complicated problems, although care and effort is needed to establish all possible sequences of actions and outcomes and the probabilities of the latter. Clearly farmers do not normally draw decision trees but it can help advisors in analysing decisions and suggesting improved choices.

Dynamic programming of farm management decisions is an application of 'control-theory'. The objective is to find an optimal policy for control of the farm system. A schematic diagram of the control process is given in Figure 5.9. The feedback represents the way the new information about the weather and other variables is used to modify the policy in multistage decision making.

Note that, as described above, the method identifies the choices which maximize expected returns. This is the optimum, only for a risk-indifferent

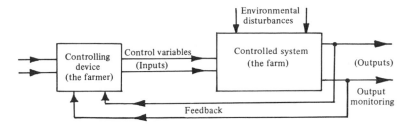

Fig. 5.9 A control system

farmer. However, the technique is easily adapted to find other possible optima, such as the maximin. Thus from Figure 5.8 we see that the maximin sequence of decisions is the same as before, since the worst possible outcome for the 'plant' strategy is £60, while that for 'do not plant' is £30. More commonly, outcomes are evaluated in terms of expected utility, rather than EMV. Since the decision-maker's expected utility takes account of his risk aversion, this is reflected in the resultant choices.

Risk-sharing institutions and policies

Risks to the individual household can be reduced by sharing. One possibility is that various individuals earning different sorts of income, which are not positively correlated, agree to pool at least part of their incomes and share the associated risks. In any season, incomes are redistributed from those who have done well to those who have done badly. The risk-pooling effect is rather like that of diversification. Provided that individual incomes are not strongly positively correlated, the income variance and the probability of disaster are reduced by sharing in this way. This principle underlies the importance of sharing within the extended family and co-operative, mutual-assistance societies. Note, however, that it is of no benefit when everyone faces the same risks so their incomes are positively correlated. Thus in a rural community if everyone's crops fail and livestock die in a drought, all suffer together.

An alternative, known as 'risk-spreading' occurs when a single risky project is shared by more than one individual. Returns are then shared but so too is the risk for each individual. This situation arises in the various forms of crop sharing and livestock caretaking widely practised in Africa. Crop sharing is common in the case of tree crops, which are tended and harvested by labourers who do not own the trees. Produce is then shared in some agreed proportion between owner and labourer. Caretaking of breeding animals, is based on sharing the progeny between owner and herdsman. In either case the risk is less if shared between the partners in the transaction.

Any method of reducing risk to the individual producer, who is risk averse, is beneficial, first, because his utility is increased by the reduction in risk, and second, because it may enable him to concentrate on maximizing expected returns, thereby increasing productivity. For these reasons policy-makers and planners may seek to reduce income variation and risk for farmers. Almost any change in the rural environment may affect the riskiness of rural life, but four types of innovation deserve special mention. First, new technologies are likely to influence production and yield uncertainty. Some high-yielding crop varieties have proved more variable than traditional ones, use of nitrogenous fertilizers on drought prone crops may increase risk, while the use of pesticides and irrigation tends to reduce risk. The need to assess the riskiness of new technology is increasingly recognized.

Second, price stabilization policies and minimum price guarantees reduce market uncertainty as do improved input supply services. However, where yields and prices are negatively correlated, so that low prices result from bumper harvests and high prices result when crops fail, product price stabilization may destabilize incomes. Provided this is not the case, market stabilization should reduce risks. Third, the provision of information to farmers regarding new technologies, market prices and new policies will also reduce uncertainty and risk.

Finally, mention should be made of crop insurance schemes as a direct method of reducing risk. Under such schemes each farmer pays an annual premium or levy on his crop, but in the event of crop failure receives compensation. The financial viability of crop insurance depends upon pooling the risks of the many farmers insured. By pooling risks, the average variance and cost of risk per farmer might be substantially reduced. But, as we have seen, this result requires that individual incomes are not positively correlated. When everyone's crop fails at the same time, the cost to the insurance agency might be very high indeed. Risks might still be spread, over a number of years, but the agency would require huge reserves to cover the occasional bad year.

An additional problem with crop insurance is 'moral hazard' which means that farmers who know their crops are insured against failure may not bother to tend them properly. This problem may be overcome by only providing insurance on an area basis; that is only to pay compensation when the district average yield falls below the critical level. The best policy for the individual producer is then to maximize his returns as he would if no insurance was offered.

For a voluntary insurance scheme, 'adverse selection' is another problem. This means that only those farmers who are most at risk would bother to join the scheme. Compensation payments per farmer would be higher than if some less risk prone farmers were insured. High premiums are required to cover high costs which further discourages farmers from joining the scheme. Many crop insurance schemes are therefore made compulsory for all growers. However, for reasons given above, most crop insurance schemes have run into financial difficulties. They may be justified none the less on welfare grounds as a means of subsidizing victims of natural disasters.

Further reading

Anderson, J. R., Dillon, J. L. & Hardaker, J. B. (1977). *Decision Analysis in Agricultural Development*, Ames, Iowa State University Press

Balcet, J. C. (1982). Adoption of Farm Technology on the Northern Nigerian Agricultural Development Project. Paper presented at the First National Seminar on the Agricultural Development Projects, Ibadan, 14–15 July, 1982

Moore, P. G. (1972). *Risk in Business Decisions*, London, Longman

Roummasset, J. A. (1976). *Rice and Risk: Decision Making among Low Income Farmers*, Amsterdam, North Holland

Roummasset, J. A., Boussard, J. M. & Singh, I. (1974). *Risk, Uncertainty and Agricultural Development*, Agricultural Development Council

Upton, M. & Casey, H. (1974). Risk and some pitfalls in the use of averages in farm planning, *Journal of Agricultural Economics*, **25**(2) 147–52

Wolgin, J. M. (1975). Resource allocation and risk: a case-study of small-holder agriculture in Kenya, *American Journal of Agricultural Economics*, **57** 622–30

PART II

Resource use

6

Natural resources

Land, wildlife and water

Agriculture and pastoralism may be viewed as ways of exploiting the natural environment. Human labour and man-made capital are employed in harvesting some of nature's bounty. As was mentioned earlier, man actually modifies his natural environment and its flora and fauna. The application of science and technology allows him to do so to an increasing extent. None the less the availability of natural resources still limits what can be harvested from a given area using a given technology.

In so far as natural resources may be destroyed by over-exploitation or misuse, choices arise between current and future consumption. Resources which are conserved for the future might otherwise have been used to increase consumption now. The analysis of such choices really comes into the realm of capital theory to be discussed in Chapter 8.

For present purposes it is convenient to assume that users aim to sustain productive capacity over time; or in other words to maintain a steady state. This means that the stock of natural resources is not allowed to deteriorate over time. Naturally, yield may vary from season to season as a result of chance climatic and other variation but a steady state implies no long-term downward (or upward) trend in the resource stock. There are three main categories of natural resources of importance to the farmer, which are:
  (i) biological, renewable resources such as forests, wildlife and natural rangeland grazing;
  (ii) non-renewable resources of land, meaning the 'original and indestructible properties of the soil'; although we know that soil fertility may be destroyed by overcropping or overgrazing and soil may be lost by erosion; and
(iii) water.

This last resource is difficult to categorize. In one sense it is renewable; both surface water ponds, drains and rivers *and* ground water aquifers are replenished by the precipitation phase of the water cycle. On the other hand, the total amount of rainfall in a given season is fixed. In this sense water is a non-renewable resource.

Water resources may be further divided into three main classes:
(i) *Drinking water* is essential for human and animal life; and may be treated as a fixed requirement or constraint. Thus the availability of drinking water influences the location of human settlements and the areas of rangeland which can be grazed. In fact some pastoralists only water their cattle every three days during dry seasons, which despite some loss of condition of the animals, does allow a larger area to be grazed. Drinking water is a variable input in this case subject to the pastoralists' choice of optimum frequency (see Nicholson 1986).
(ii) *Soil moisture* is essential for crop growth, and its availability is an important influence on the productivity of the soil. For many purposes it may simply be treated as a characteristic of the land.
(iii) *Irrigation* generally yields substantial increases in productivity per hectare, which may justify considerable expenditure on pumping, storage and transport. In this way it becomes possible to control the allocation of water between crops and between time periods. In principle, the optimal allocation of water requires that the value of the marginal product per cubic metre is the same in all uses and equal to the supply cost, after having made due allowance for risk (see previous four chapters). Reduction of the risk of losses due to drought may be a major benefit of irrigation. In practice there are many special problems associated with the control and distribution of

irrigation water. Those interested in pursuing the subject further might start by reading Carruthers & Clark (1981).

The exploitation of biological resources

The following analysis may be applied to any biological, renewable resource, be it a fishery, a herd of wild game, a forest harvested for firewood and building materials or natural rangeland.

Provided that the rate of harvest offtake does not exceed the natural growth rate, the stock will not be depleted. A steady-state ecological equilibrium may be maintained. If, on the other hand, over-exploitation occurs, the natural resource stock will be depleted or even destroyed.

To illustrate let us first consider the growth of a biological population or stock of plant biomass, which is not harvested. The growth curve is usually S-shaped as shown in Figure 6.1 since growth is slow when the stock is sparsely distributed, increases exponentially as the stock grows, but gradually slows down as the stock increases further, due to 'environmental resistance'. Environmental resistance may be due to overcrowding, pressure on food supplies for animals or shading effects for plants. Growth ceases

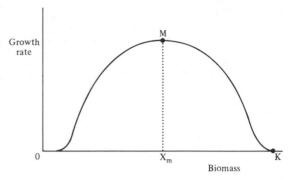

Fig. 6.2 The effect of biomass on growth rate

once the environmental 'carrying capacity' is reached. This is a somewhat simplistic approach in not distinguishing between reproductive and other phases in the population and ignoring interactions between different species, but it will serve our purpose.

Figure 6.2 which shows the relationship between the total biomass and its rate of growth is easily derived from Figure 6.1. In reality the curve could not originate at zero, because some minimum initial stock is necessary to reproduce and grow. However, the important features to note are M, the point of maximum growth, at stock OX_m and OK, the natural equilibrium stock. Without human exploitation the stock would stabilize at this maximum level.

Now let us assume that humans harvest some of the resources by fishing, hunting, felling trees or putting cattle to graze on the rangeland. The effects of three different harvest rates are shown in Figure 6.3. First consider an annual harvest offtake of OH_1 on range-

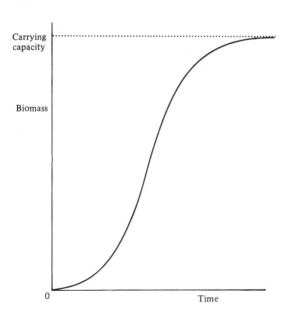

Fig. 6.1 A biological growth curve

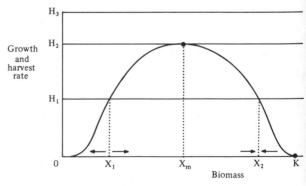

Fig. 6.3 The effects of different harvest rates

land where it would represent a relatively low stocking rate. A steady-state equilibrium would result with a biomass of OX_2. This rate of exploitation could be sustained indefinitely, barring disaster, because the natural growth rate is just sufficient to replace the harvested offtake. Furthermore, this is a stable equilibrium, which means that if the balance is disturbed it is automatically restored. Thus, if as a result of a poor growing season, the biomass is reduced below OX_2, the natural growth rate will increase. Since the offtake rate remains constant the total biomass must increase, thus restoring the stock to level OX_2. Similarly a chance increase in the biomass above this level would be self-eliminating.

A steady-state equilibrium also exists at a much lower biomass, a smaller population of wild animals or a more denuded landscape, at OX_1. However, this is an unstable equilibrium. In a bad year, if natural growth fell below the harvest rate, further depletion would occur. A reduction in harvested offtake would be needed to arrest disaster. In a good year, on the other hand, the stock would increase and with it the natural growth and regeneration rate. This self-sustaining growth would continue until the stable equilibrium is reached at OX_2.

A harvest offtake of OH_2 represents the 'maximum sustainable yield' (MSY). It is the maximum possible offtake which is a steady-state equilibrium. By contrast a harvest rate of OH_3 cannot be sustained. It may be possible to harvest this yield for a short period, but only at the expense of depleting the resource stock. This process of over-exploitation or resource mining, if continued, must ultimately result in total destruction of the resource.

Now it might appear that the maximum sustainable yield is the optimal rate of exploitation, since biological resources are free gifts of nature. However, there are two main reasons why this might not be the case. First, there are risks in trying to operate at this level, since even a slight fall in natural growth during a poor season must result in a reduction of the total biomass below OX_m. This in turn means that growth will fall below MSY in future seasons. If the level of offtake is not reduced accordingly continual depletion is inevitable. Risk averse users of the natural resource are likely to aim for a larger steady-state biomass than OX_m.

Second, costs are incurred in harvesting natural resources; in hunting, fishing, collecting firewood or keeping livestock. Furthermore, the marginal and average returns per unit of cost or effort are likely to diminish as the activity is increased. To see why we may refer again to Figure 6.3. With no exploitation the total biomass is given by OK, but as the harvesting effort is increased first to OH_1 and then to OH_2 the biomass is reduced through OX_2 to OX_m. Thus harvesting takes longer. Returns to hunting or fishing fall as the population of game is diminished. The same is true of wood-gathering as the forest is thinned out, or of keeping more livestock on rangelands as the grass cover is reduced. Similar arguments might even be applied to ground water, since the yields of wells and boreholes may diminish as the water table is drawn down. In all these instances the situation of diminishing returns to effort is similar to that for weeding labour analysed in Chapter 2 (see Figures 2.1 to 2.3). Now it is quite feasible that the economic optimum rate of harvest lies below the MSY, provided that the product value is not too high in relation to the effort required. Increased human population pressure may raise the scarcity value of natural resources and cause a higher rate of exploitation.

To summarize this section we may conclude that

(i) provided the rate of offtake does not exceed MSY and the biomass does not fall below OX_m, a steady-state equilibrium harvest can be maintained, and

(ii) diminishing returns are likely to result from increased harvesting effort and the optimum rate may lie below the MSY.

Intensity of land use

In those parts of Africa where there is still some unexploited virgin bush, land use per household can be varied. The economic optimum policy is to operate an 'extensive' system with a large land : labour ratio and hence low labour inputs and yields per hectare, for instance under long-fallow shifting cultivation. This is illustrated using the marginal and average product curves for labour shown in Figure 6.4. The average product of labour is maximized at point B so this is the optimum. Labour input per hectare is limited to OA and total product per hectare is limited

58   *Resource use*

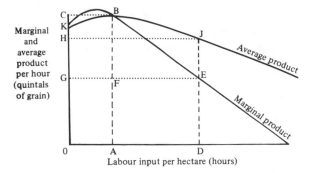

Fig. 6.4 Returns to labour under different intensities of
land use

Table 6.1. *Labour input in shifting and fallow systems,
Kiberege Strip, Kilombero Valley, Tanzania*

	Shifting cultivation on the escarpment	Fallow farming in the valley
Cultivated area as a percentage of total cultivated and fallow (cropping index)	15	55
Total area cultivated and fallow (ha/ME)[a]	4.0	1.7
Land productivity (kg rice/ha)	252.5	654
Labour input (man–days/ha)	41	136
Labour productivity (kg rice/man–day)	6.2	4.8

[a] ME = man-equivalent.
After Baum (1967)

to OABC. Population growth can be accommodated simply by increasing the area of land use to scale (see Chapter 3). In these circumstances land is truly a free resource.

Once all available land is in use, as a part of a crop rotation, increased labour inputs must mean intensifying the system of land use to produce more per hectare. The effects may be illustrated in Figure 6.4. When labour input per hectare is increased from OA to OD, marginal and average products per hour of work fall to OG and OH respectively. However, the total product per hectare is increased from area OABC to area ODJH. Alternatively, measuring total product by the area under the marginal product curve, the increase is given by area ABED. Thus returns per hour of labour fall with increased labour use per hectare, while returns per hectare of land rise.

Although there is some scope for varying the intensity of land use within a given system, pastoralists can vary the stocking density of the rangeland for instance, bigger changes can be accommodated by changing the system. A change from pastoralism to arable cultivation, generally allows increased intensity of land use. Instead of relying on natural regrowth, the cultivator actually harvests the entire crop biomass (OK in Figure 6.3) each season and re-establishes the crop in the following season. This requires more labour, but produces more yield per hectare.

One way of describing the intensity of cultivation is in terms of the frequency with which the land is cropped. This is measured by the 'cropping index', which is the number of crop seasons as a percentage

of the total rotational cycle including fallow. On this basis arable cropping systems may be classified, according to the intensity of land use, along the following lines:

1. Natural fallow systems
1.1 Shifting cultivation usually with forest fallows and a cropping index of below thirty-three
1.2 Fallow systems, usually involving bush or savannah grass fallows, with a cropping index of between thirty-three and sixty-six
2. Permanent cultivation systems
2.1 Annual cropping under which land is only fallow between seasons or at most one year in three. The cropping index lies between sixty-six and one hundred
2.2 Multi-cropping with each plot bearing two or more crops per year, giving an index of over one hundred.

From the analysis of Figure 6.4 we might expect increases in returns per hectare and decreases in returns per man–day of labour with increases in the cropping index. Many field studies support these conclusions; such as the example given in Table 6.1 (Baum, 1967). Widely quoted examples of intensive, high labour input systems are found in the close settled zones around major cities in Northern Nigeria, on the island of Ukara in Lake Victoria and among coffee farmers on the slopes of Mount Kilimanjaro.

Some changes of system clearly involve the use of more capital. Examples are the establishment of permanent crops, or the construction of irrigation works and terraces to grow flood rice. The same is true where ley farming, the use of rotational grassland for feeding livestock in a mixed farming system, involves the purchase of livestock, buildings and equipment besides the cost of establishing the ley. In each of these cases, the use of capital may bring about an increase in output per hectare, possibly increased returns to labour too.

Increasing intensity of land use, without a change of system to restore soil nutrients, may result in declining fertility and ultimately loss of soil structure and soil erosion. Studies in high population density areas of Eastern Nigeria show cassava yields falling dramatically from 10.8 tons to 2 tons per hectare as the length of fallow is reduced from 5.3 years to 1.4 years (see Lagemann, 1977). The situation is rather like that of exceeding the MSY for a biological resource. If continued long enough the resource may be destroyed. These problems will be taken up again in discussing property rights and tenure.

Variations in returns to land

In theory, the returns to land and labour can be separated according to their marginal productivities. For simplicity we ignore returns to capital and assume that land and labour are the only inputs. Referring again to Figure 6.4, it may be argued that the return to labour, which is the marginal product times the labour input, changes from area OABC to area ODEG with increased intensity. But, for the extensive system, the return to labour OABC is the same as the total product. There is no separate return to land; it is zero. However, with increased intensity, the surplus over the labour cost, area GEBK, (which equals GEJH) may be viewed as a return to land, or 'rent'.

Alternatively, if labour could be hired at a wage of OG, then the economic optimum level of employment per hectare would be OD, so the amount

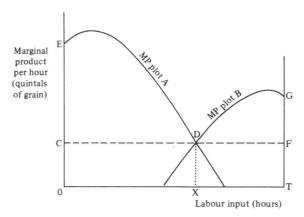

Fig. 6.5 Efficient labour allocation between two plots of land

ODEG would be paid out as wages. The area above this, which we have identified as rent, is the surplus remaining for the land holder. In practice, where both land and labour are provided by the farm family, factor shares cannot be distinguished in this way.

However, in order to make efficient allocation decisions it is important to identify opportunity costs as we have seen in Chapter 4. An alternative view of the allocation problem is given in Figure 6.5. This shows the marginal product curves for labour applied to two different one hectare plots of land. Note that the marginal product curve for Plot B is drawn back to front, since labour applied to this plot is measured from right to left, starting from point T. The total labour hours available for cultivating the two plots are measured by OT. Given that no labour is wasted, an increase in employment on one plot must mean less labour employed on the other. Plot A is more fertile than Plot B so the marginal product of labour is higher on A for a given intensity of labour use.

Several useful conclusions may now be drawn.

(i) The optimum, most productive, allocation of labour is found when the marginal product is the same for both plots, at point D, where OX hours are allocated to Plot A and XT hours to Plot B. This is simply an application of the equi-marginal returns principle; that scarce resources should be allocated so that marginal returns are the same in all uses. It is readily seen in Figure 6.5 that an increase in labour used on plot A beyond OX would result in a smaller gain in yield than would be lost from the reduced labour input on Plot B. A similar net loss would occur if labour was transferred to Plot B, away from the optimum allocation.

(ii) At the optimum allocation, more labour is used on the more fertile plot (A). In short the more fertile plot is cultivated more intensively.

(iii) This marginal product of labour, which is the same on the two plots, represents the opportunity cost of labour. If this hourly cost is applied to the total labour input OT, the area OTFC represents the return to labour. The areas CDE and DFG, above this, represent returns to land.

(iv) The return or rent per hectare of the more fertile plot (area CDE) is greater than that of the less fertile plot (area DFG).

Major differences in land quality are often found within the area of a single farm and cropping systems are organized accordingly. The crops chosen for fertile valley bottoms, such as rice, sugar cane and vegetables, probably yield a larger surplus per hectare, which means that they are more intensive crops than upland sorghum or millet for instance. Similarly yam is a more intensive crop than cassava. Such comparisons require that money, or some other common unit of value, is used to compare the returns from different products. However, for present purposes it is useful to remind ourselves, by using a physical measure of output, that land allocation decisions are made even by pure subsistence farmers.

In reality there are added complications in that there may be more than two classes of land and it is not neatly divided up into one hectare plots. However, there must be a limit on the amount of each class of land available. Thus once all the top-quality land is in use, as the intensity is increased, a stage will be reached when labour can be used more productively on poorer quality land. This is illustrated in Figure 6.6, which shows the marginal product curves for labour on three different classes of land. Once OA hours of labour are used on the most productive land, further labour would be used more productively by extending on to second-quality land.

As yet more labour is used, the marginal product falls on second-quality land until eventually (when AC hours are worked on this land) it becomes worthwhile to start cultivating even poorer, third-class

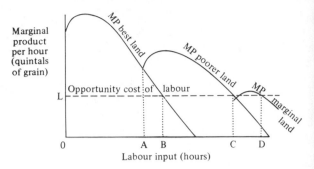

Fig. 6.6 The intensive and extensive margins of cultivation

land. There are two important effects. One is that as the marginal product of labour falls on second quality land, that is the opportunity cost falls, the intensity of using top-quality land increases from OA hours of work to OB hours. This is known as the effect on the 'intensive margin', when costs fall or returns rise.

The second effect is that the area of cultivation is extended on to poorer quality land. Note, however, that poorest quality land is freely available and sufficient is used, with low labour inputs per hectare, for the labour cost to account for the total product. There is no surplus over labour cost on the poorest land, which is therefore known as 'marginal land'. It is only just worth cultivating. The return to labour on this marginal land determines the opportunity cost. A surplus is produced over and above this on better-quality land. These ideas may become clearer when we consider the effect of location which also influences the surplus produced per hectare.

Location

Most farm families maintain a compound or kitchen garden close to the house, which is fertilized with ashes and household refuse and continuously cropped. Land further from the home is used less intensively under a rotational bush fallow system for instance. This may reflect a conscious choice to build the settlement on the most fertile soil in the area. However, even if all the land was of equal fertility, that near the home would be cultivated more intensively because of its location. Time is lost walking to and from more remote plots, so the surplus over labour costs must be lower. As we have seen from

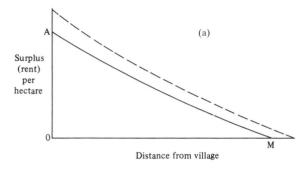

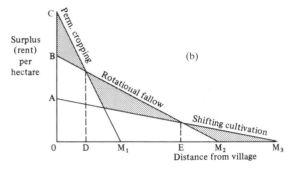

Fig. 6.7 The influence of location on land use, (a) The distance–decay function, (b) Alternative forms of land use

Figure 6.5, this means the remote plots will be cultivated less intensively.

Figure 6.7(a) shows a continuous decline in the surplus produced per hectare with increasing distance. At distance OM, there is no surplus, so land is marginal in this vicinity because of its remoteness from the village. Beyond this, there may be unused land, which is sub-marginal or not worth using. Assuming there are no obstacles, so that access is equally easy in all directions, the margin of cultivation may be represented by a circle around the village. Note that a general increase in productivity represented by the broken line in Figure 6.7(a) would cause both an increase in intensity on land already cultivated (the intensive margin) *and* an extension of the area (the extensive margin).

This analysis can be pursued a little further. Consider first, the continuous cultivation of vegetables. As an intensive form of land use it requires frequent visits throughout the year. Time lost walking to and from the plot would be serious if the plot was far from the village. In short the surplus per hectare would decline rather sharply with increased distance, as shown by the line CM_1 in Figure 6.7(b).

For less intensive rotational cropping requiring less frequent visits, the surplus is lower at or near the village, but it declines more slowly with distance as shown by the line BM_2. Finally for shifting cultivation, the return per hectare is low but it falls even more slowly as shown by line AM_3. On this basis, we can see that the most productive use of the nearest land is intensive vegetable growing, up to distance OD from the village. From distance OD to OE, rotational bush fallowing is more profitable. Beyond this, and even on land which is sub-marginal for continuous cropping, shifting cultivation may be justified. The shaded areas represent the differential rent or surplus earned over the next best alternative form of land use.

Thus location has an important influence on farm systems. A pattern of concentric rings of different forms of land-use has often been observed around villages as shown in Figure 6.8 (after Pelissier, 1966).

Common rights and conservation

Natural resources are generally treated as the common property of the whole of society. In the well-known words of a Nigerian chief 'land belongs to a vast family of which many are dead, a few are living, and countless numbers are still unborn' (Elias, 1962). Of course, access is not open to anyone; it is generally limited to members of the nation, lineage or village society. Furthermore, user rights to cultivated land are allocated to individual households. However, in much of Africa no rents are paid for land use while access to forests and rangelands is open.

It has been argued that open access, to rangelands for instance, necessarily leads to their destruction and ruin. This is the so-called 'tragedy of the commons', (Hardin, 1968). In fact, this argument is false. As we have seen, given that there are diminishing marginal returns to increased effort in exploiting natural resources, there is clearly an economic limit to the rate of exploitation. Indeed, the economic optimum rate may lie below the MSY for biological resources or, in the case of cultivated land, the optimum intensity may lie below the level where

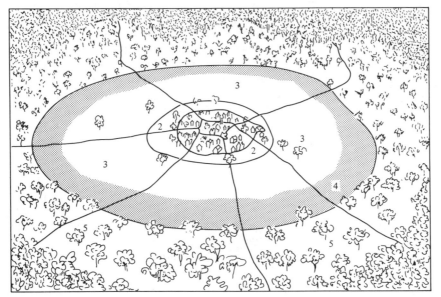

Fig. 6.8 Spatial organization of land use in N'Gayene, Senegal (from Pelissier 1966, p. 474). 1 Houses and gardens. 2 Permanent cultivation. 3 Intensive fallow systems. 4 Intensive shifting cultivation. 5 Bush and extensive shifting cultivation

fertility decline occurs. Alternatively, measures for fertility maintenance and conservation may be incorporated in the optimal policy. In these cases a steady-state equilibrium may be maintained.

The only germ of truth in the 'tragedy of the commons' is that open access may result in a higher rate of exploitation than would occur under private ownership. This is illustrated in Figure 6.9, which shows the marginal and average product curves to labour or effort applied to natural resources. As already emphasized, the economic optimum (under private ownership) occurs where the marginal product is equal to the unit cost of effort (level OA). However, at this point, there is a surplus or 'rent' since the average product exceeds the average cost. Under open access, where no rent is paid, there is an incentive to exploit the resource further and increase the effort to the point where average product equals the unit cost (level OB). There are differences of opinion as to how this effect should be interpreted. To the environmentalist it represents over-exploitation, but to the developer it may be seen as fuller and more productive use of available resources. The interpretation should depend upon whether the resource is being depleted.

Natural resources may be depleted either as a result of chance, droughts, floods or disease outbreaks or because of a conscious decision to do so. In the latter case the users place a higher value on their current consumption than on the welfare of future generations; they discount the future. This might arise as a result of (i) opportunity, such as high prices for wild game, or (ii) need, because of rapid human population growth. In these circumstances, open access might hasten the rate of resource depletion.

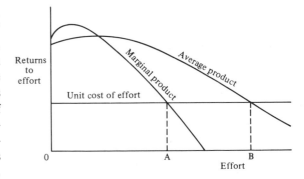

Fig. 6.9 Exploitation of natural resources under open access

Social costs are incurred, in terms of loss of future production, which are not necessarily felt by current users.

Chance effects are important since much uncertainty surrounds the use of natural resources. It is difficult for farmers, and scientists, to monitor and evaluate when over-exploitation and resource depletion is occurring. Random variations around a steady-state equilibrium may be either self-correcting *or* self-perpetrating. This implies that a safety margin is desirable; the planned intensity of exploitation should be below the expected MSY.

Where there are risks of depletion the Government may wish to impose some control on the exploitation of natural resources. One or other of the following alternative approaches may be appropriate depending on the circumstances:

(i) Privatization; for instance, by enclosing, fencing and issuing titles to grazing land, or allocating and registering private water rights, should in theory, provide the necessary incentive to operate at the point where marginal product equals cost of effort. Furthermore, given security of tenure, it should also provide an incentive to conserve the resource. However, over-exploitation may still occur for the reasons given above, social inequalities may develop under private ownership and the concept may be socially unacceptable.

(ii) Compulsion, of which there are two main forms (a) imposition of quotas, on the number of livestock grazed or the amount of game, firewood or water taken annually and (b) taxes per unit harvested or per head of livestock. In theory, quotas or taxes can be adjusted so as to maintain any desired rate of natural resource use. In practice, adjustment may be difficult because of the uncertainties mentioned above, and there are costs of monitoring and surveillance. Government monopoly of natural resources, and charges for their use may be socially unacceptable.

(iii) Persuasion and assistance. Many instances have been quoted (eg Richards, 1985) of cultivators and pastoralists adopting measures for soil and water conservation. Little persuasion may be needed, where farmers are aware of the risks.

However, an environmental monitoring and information service might enable farmers and pastoralists to adjust better their rates of usage of natural resources. Programmes for reducing population growth and raising rural incomes will reduce the pressure on current use of natural resources. In some cases direct assistance may be needed for terracing, reforestation, land reclamation and other conservation measures. As with all spending on rural development the costs must be weighed against the future benefits.

Land tenure

The system of land tenure is the set of laws and customs which establish rights and duties relating to land use. From the national point of view, the two main requirements of a system of land tenure are firstly, that it should lead to the most productive distribution of land among potential users and potential uses; and secondly, that it should provide sufficient security of tenure to justify measures to maintain or improve the productivity of farms. To some extent these two requirements are in conflict, since the greater the security of tenure, the more difficult it becomes for efficient farmers to obtain control of land occupied by the less efficient. Most of the customary tenures in Africa also emphasize a third feature: the right of each family to some share in the natural resources belonging to the community, the lineage or village. Sale or lease to strangers may be prohibited but, subject to these overriding controls, individuals belonging to the community have the right either to use particular pieces of land which may have been inherited, or to occupy plots under shifting cultivation.

African patterns of land tenure have proved to be flexible in allowing for differences in the needs of families and for changing circumstances. It has generally been possible for the more energetic or capable farmers to obtain extra land, either from that controlled by the community and still unused, or by way of pledge or loan from other families, who have more land than they can use. Again, the introduction of tree crops in both East and West Africa led to extension of individual rights in land to cover the prolonged period of occupation and to allow the pledging and even sale of cocoa or coffee plots which a man has built up by his own work. In such cases systems of land tenure may evolve naturally through increasing individualization to the stage of a commercial market in land, where plots can be bought and sold like other commodities.

However, the existing pattern of land tenure may not evolve quickly enough to avoid acting as a constraint upon agricultural changes which are urgently required in view of the rising populations and the expansion of export crop production. In the first place the loan or pledging of land, based on close personal relationships, provides neither the security of tenure nor the incentive for efficient farming by the occupiers. Secondly, the sale of land becomes associated with lengthy and costly legal disputes regarding the true ownership, particularly if it is sold to strangers from outside the community. Hence, with the development of a market in land, it becomes important to establish individual rights on a legal basis by survey and registration of ownership. Thirdly, the system of inheritance may lead to excessive fragmentation, which causes time loss in walking to and from isolated plots and creates problems for any kind of mechanization. Land consolidation and legal control of subdivision may contribute to increased agricultural productivity.

Where the individual owner has the sole right to use a piece of land or to dispose of it as, and to whom, he wishes it is known as freehold tenure. It is widely believed, particularly in East Africa, that individual freehold tenure is a highly desirable, even an essential, component of any agricultural development programme. The pride of ownership and the security offered to the farm family by this form of tenure are thought to encourage long-term improvement and conservation of the land and associated water resources. Furthermore privately owned land can be offered as security for loans, to be given up to the lender in the event of failure to repay the loan; that is it can be mortgaged, thus enabling small farmers to raise money for farm improvement. However, this argument is of doubtful validity since few banks and commercial moneylenders are prepared to accept land as security for loans because of the practical and political difficulties of removing farmers from their

land if they fail to repay their loans. It is also argued that the market in freehold land encourages the able and industrious farmers to expand production by buying the land of the less successful thus encouraging the development of a commercial attitude to farming.

This last feature of private land-ownership has its disadvantages in the potential for increasing inequality of land holdings and the development of a rural landless labouring class, as is occurring in Kenya (Hunt, 1984). Where, in addition, the leasing of land is prevalent, class distinctions may be established between landlords and tenants. This was the case in Ethiopia before the 1974 revolution and subsequent land reform. However, the alternative to private land ownership, of communal farming of land held in common has had limited success there and in Tanzania. Cultivation of individual family holdings still continues in both countries.

In those countries of Eastern and Southern Africa where large-scale commercial farms were established under colonial governments, some of these have survived alongside much smaller semi-subsistence holdings. Redistribution of land was one of the objectives of the independence movement, and, as we have seen, production per hectare may well be higher on small farms. Both equity and efficiency objectives may be served by the subdivision and resettlement of the large-scale farms. However, substantial costs may be involved in compensating the occupiers and resettling smallholders, while problems may arise in maintaining urban food supplies since most of the marketed surplus comes from commercial farms.

Leasehold

Forms of leasehold tenure, met with in African countries include formal fixed-rent tenancies, crop-sharing agreements and pledging in which case the payment is in the form of a loan. Where there is competition between potential tenants, with the lease being awarded to the highest bidder, the rent paid will equal the total surplus over operating costs. More precisely, the rent is the same as the marginal product per hectare. Arguably the payment of rent promotes efficiency. The tenant must operate at the economic optimum intensity in order to afford the rent. Any increase in crop prices or productivity will cause the demand for rented land to rise and so will the rent. This effect does not apply, of course, where rents are fixed below the free market rate and eviction of tenants is forbidden as was the case under the *Mailo* system introduced in Buganda. There, tenants were able to retain some of the surplus and accumulate enough to buy the land. This reminds us that even when no rents are paid, the producer still has an incentive to use land productively. The supposed advantage of charging rents must be set against the lack of security of leasehold tenure and the income distributional impact of a landlord–tenant system. Note, however, that on many irrigation and settlement schemes the state is the landlord.

It used to be argued that crop-sharing agreements are inefficient. The tenant, since he only keeps a share of the produce, has less incentive to maximize the surplus over production costs. He would only apply labour up to the point where his share of the marginal product equals the wage rate. It is now recognized that costs per hour of labour are not fixed, and the tenant might well aim to maximize total production rather than the surplus over some notional labour cost. In any case, if this system really were less efficient, it would presumably be replaced by a fixed-rent system. Share-cropping is really, as the name suggests, a system for sharing productive resources, the associated risks and the benefits.

Increasing output per hectare

Although there are high land/man ratios in Africa, we must not assume land is unlimited. Much is under fallow in rotational systems, and effectively in use. Furthermore, as we have seen, better quality and better located land is likely to be fully used before production is extended onto poorer and more remote land. Thus the area of more productive land is a constraint on output. New technology which increases output per hectare will bring about growth at both the intensive and the extensive margins. Examples of such technology are new high-yielding crop varieties, fertilizers and pesticides. For instance, introduction of short-season maize may allow two crops to be taken in place of one each year and production to be extended to drier areas. Systems of agro-forestry, with soil mulching, may help maintain fertility and increase intensity of land use, thereby also increasing output per hectare. Although less obvious, increases in livestock productivity through improved health, nutrition and breeding also raise output per hectare.

Such innovations are sometimes described as 'land-saving', in the sense that less land is needed in total to produce a given output. However, strictly speaking, this description which implies that capital and labour are substituted for land, may be misleading. In general, new technology which raises crop and animal yields is neutral, at least between land and labour so that yields per hectare *and* per man–day are raised together. Admittedly more harvest labour may be needed for a larger crop yield, and mulching requires extra labour. To set against this, any extension of the feasible number of years of continuous cultivation, reduces the need for bush-clearing labour. Overall, labour productivity is likely to be raised.

Furthermore, although costs may be involved, there are no large items of capital (except perhaps for pesticide sprayers). Thus most of these innovations are scale neutral. They can be introduced equally easily on small or large farms. In practice, large farmers may have easier access to the new inputs and credit to buy them, so they may adopt new technology more quickly. Where there are effective input supply systems such innovations can spread rapidly even among small farmers. One well-documented example is the introduction of hybrid maize to Western Kenya, where the innovation spread to most farms within two or three years (Gerhart, 1975).

To derive the full potential benefit of introducing modern varieties may require application of fertilizers and use of pesticides, so a package of these innovations is likely to be more productive than one on its own. Furthermore the impact may be felt throughout the farm household system. The introduction of hybrid maize for instance may cause a shift away from mixed cropping, more land devoted to maize for cash sale, or less land and labour devoted to maize if a target quantity is needed for subsistence, and the purchase of seed rather than retaining it from year to year. Thus careful testing and evaluation of the overall impact is needed before innovations are launched. This must include an assessment of the risks involved. Generally speaking, yield increasing innovations also increase the total inter-seasonal yield variation. The *proportionate* variation (coefficient of variation) may well decline. Some new varieties have been unacceptable to farmers because of the appearance, processing characteristics, cooking qualities or taste. Clearly, such factors are as important as the economic impact on the farming system in determining the acceptability of the innovation.

Further reading

Adegboye, R. O. (1977). 'Land Tenure' in *Food Crops of the Lowland Tropic*, eds Leakey, C. L. A. & Wills, J. B., Oxford University Press

Baum, E. (1967). 'Land Use in the Kilombero Valley' in *Smallholder Farming and Smallholder Development in Tanzania*, ed. Ruthenberg, H., Afrika-Studien, No 24, IFO Institute, Munich

Boserup, E. (1965). *The Conditions of Agricultural Growth*, London, George Allen & Unwin

Carruthers, I. & Clark, C. (1981). *The Economics of Irrigation*, Liverpool University Press

Cohen, J. M. (1980). 'Land Tenure and Rural Development in Africa' in *Agricultural Development in Africa: Issues of Public Policy*, eds Bates, R. H. & Lofchie, M. F. New York, Praeger

Dasgupta, P. (1982). *The Control of Resources*. Oxford, Blackwell

Elias, T. O. (1962). *Nigerian Land Law and Custom*, London, Routledge & Kegan Paul

Gerhart, J. (1975). *The Diffusion of Hybrid Maize in Western Kenya*, CIMMYT, Mexico

Guillard, J. (1965). *Golonpoui, Analyse des Conditions de Modernisation d'un Village du Nord-Cameroun*, Mouton, Paris

Hardin, G. (1968). The tragedy of the commons, *Science*, **162**, 1243

Hartwick, J. M. & Olewiler, N. D. (1986). *The Economics of Natural Resource Use*, New York, Harper & Row

Hunt, D. (1984). *The Impending Crisis in Kenya: The Case for Land Reform*, Aldershot, Gower

Lagemann, J. (1977). *Traditional African Farming Systems in Eastern Nigeria: An Analysis of Reaction to Increasing Population Density*, Munich, West Germany, Weltforum

Nicholson, M. (1986). How water shortage can benefit pastoral society, *ILCA Newsletter* **5**(2)A

Pelissier, P. (1966). *Les Paysans du Senegal*, Imprimerie Fabregue Saint-Yrieux

Richards, P. (1985). *Indigenous Agricultural Revolution: Ecology and Food Production in West Africa*, London, Hutchinson

7

Labour

The family labour force

In much of rural Africa, access to labour, rather than land, is the basis of economic and political power. This reflects the relative sparseness of population and absence of labour-saving machinery, as a result of which the labour available for critical tasks such as planting and weeding crops or watering livestock is an effective constraint on production. Although some labour is hired, the core of the farm workforce is made up of family members. The size of farm, or of livestock herd among pastoralists, depends upon the number of active family members. So long as the marginal product per person is greater than the cost of extra subsistence a large family is desirable, among other reasons, because a man can thereby increase his wealth and power.

The amount of labour used, that is the actual labour input over a given period on a particular farm, depends upon the family structure, the number of hours worked and the rate of working per hour. A common trend in much of Africa is that the traditional extended or joint family unit, consisting of more than one married man plus dependants, is breaking up into nuclear, simple family units, each consisting of one married man plus dependants within the same compound. In some communities wives may live in separate compounds, and crop separate plots of land. There is increasing individualism in farming.

None the less, some farm work is still done by age-sets and lineage groups on a co-operative or reciprocal basis. Such group activities may be essential for tasks such as land clearing or raising water from deep wells by hand, but they generally add to the social life of the community rather than to the local supply of labour.

Within the family, there is generally some division of labour, many tasks being traditionally considered as age- and sex-specific. Most domestic work, such as fetching wood and water, cooking and cleaning, is the responsibility of women. Other than in Muslim communities where they are kept in *Purdah*, women also contribute to farm work, especially food-crop harvesting and processing, and the care of small stock. Men generally carry out heavier tasks such as bush clearing and land preparation and take sole responsibility for the cultivation of cash crops. There is some debate over whether this allocation of responsibilities is just or fair. However, our main concern is simply to emphasize that labour is not a single homogeneous input but consists of various different age- and sex-groups. Efficiency is improved by division of labour if each group specializes in those tasks they perform best. In any case, there is growing evidence that sex roles change over time in the process of development.

Work inputs are usually measured in man–hours or man–days, which is simply the product of the number employed and the time worked by each. It is therefore assumed that the labour of one man for one hundred hours is equivalent to the labour of one hundred men for one hour or five men for twenty hours. This assumption ignores the possible advantages of team work for some operations, but will serve for most purposes. Conversion factors have been suggested for representing work done by women and children in man–hour equivalents (eg women two-thirds and children one-third) but these are essentially arbitrary methods for aggregating different types of labour into a single total.

Farm surveys throughout Africa and over many decades have consistently shown relatively low labour inputs in agriculture when compared with farmers in other parts of the world or industrial workers in Africa. Typically, adult males spend between 500 and 1500 hours per year on their farms (See Cleave, 1974, Byerlee *et al.*, 1976). This may be compared with 2500 to 3000 hours worked in Egypt and other regions where irrigated agriculture is prevalent, and in urban industry. In general, the hours worked per year in farming tend to be lower in arid regions because of the short growing season. The production of permanent crops in more humid areas may increase the work load.

Even in non-Muslim areas, women generally work fewer hours on the farm than men. However, when household work is included the total labour input of women exceeds that of men. Children contribute a lot less time to family activities than adults. There are various possible contributory factors to the low labour input, despite the relative scarcity of labour. First, a limit on work capacity may be imposed by climate, health and nutritional status of family members. High temperatures, especially in the middle of the day, may restrict working hours, while ill health and undernutrition may prevent work altogether. By comparison, work in rice paddies is less debilitating because of the cooling effect of the water. Secondly, studies of labour inputs in farming may have omitted the time spent walking to the fields, and to markets. Inclusion of this time might make a significant difference. The three remaining causes, namely the seasonality of farm work, the importance of so-called 'leisure activities' and off-farm work will be discussed in more detail.

The seasonality of farm work

One factor contributing to the impression that labour is underemployed in African agriculture is the marked seasonality of farm work. As a result of the seasonal variation in rainfall and crop growth, busy periods or 'work peaks' alternate with slack periods or 'work troughs'. An example of a labour profile showing these peaks and troughs is given in Figure 7.1, which represents the monthly labour use on a farm in South Eastern Nigeria. Note that the decision to divide the year's labour input into monthly periods is essentially arbitrary; the pattern of peaks and

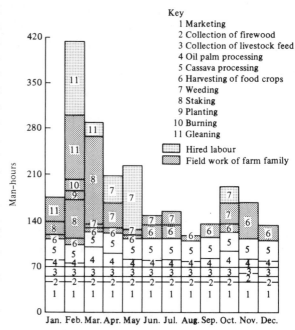

Fig. 7.1 Distribution of farm activities over one year. *Source*: Lagemann, J., Flinn, J. C., Okigbo, B. N. and Moorman, F. R.: p. 23

troughs might appear different if different (say fortnightly) periods were used. For this farm, located in the semi-humid zone, the main work peak involves clearing bush, planting and staking yams at the start of the main rains. The seasonal variation in crop work is generally even more marked in semi-arid regions with only one cropping season. Under irrigated agriculture, on the other hand, labour requirements are more evenly spread throughout the year. For livestock keepers work peaks may occur in the dry season if fodder is cut and carried, or water is raised from wells. Work peaks occur because critical tasks such as planting, weeding and harvesting are closely related to the seasons and must be completed within a limited period of time.

Delays generally cause loss of yield, so the man–hours needed to finish the task are concentrated into a peak period. Other household tasks, particularly maintenance and repair work, allow greater flexibility of timing. Unfortunately, the farm labour supply can rarely be varied from day to day so as to match requirements. The temporary hire of seasonal labour would reduce the pressure at peak periods for an individual family, but for the wider rural society the seasonal peaks and troughs of local employment remain. In parts of West Africa, seasonal labour migration between different agro-ecological zones provides a partial solution (see Swindell, 1985).

For the family workforce and regular hired workers, the supply of effort is relatively constant throughout the year. It is therefore probable that the farmer will either have less labour than he wants on the farm at work peaks, or more than he wants at slack times, or something of both. Some seasonal unemployment or underemployment is almost inevitable in agriculture. This is illustrated in Figure 7.1 where in most months, other than February and March, relatively few hours are worked by family members.

In effect this means that the marginal product and hence the opportunity cost per hour of labour varies from season to season. Labour availability at peak periods may be a critical limiting constraint on production, despite underemployment at slack periods. At work peaks an extra hour of labour would yield a considerable increase in total product, either because of more timely completion of the job, and hence higher yields, or because a larger area of crops may be cultivated. At slack periods, the marginal product of an extra hour of labour may be zero. There is no single meaningful value for the marginal product or opportunity cost of labour which applies throughout the year. The seasonal variation in the opportunity cost of labour creates considerable complexity in farm planning and the management of labour.

Seasonal labour allocation

Although there are seasonal peaks of activity associated with all crops, they do not necessarily coincide. The labour profile for yams, which is the pattern of monthly labour requirements per hectare over the year, is different from that for maize. How then does the farmer decide to allocate his labour between them? Yams may yield a higher return per man–hour in one month, while maize yields a higher return per man–hour of labour in a different month. Yet the areas planted to yams and maize cannot be varied from month to month according to their

returns per hour of labour. Thus, it is not possible to allocate labour so as to earn equal marginal returns from all crops in each and every month.

If labour can be used productively more evenly throughout the year, the total product of the family labour force will be increased. The wellbeing of the family is thereby improved. By the same token, the production per unit of peak-period labour is increased. However, it should be noted that the *total* work load over the year may also be increased so much that the average product per unit of annual labour input is reduced. This is likely to be a less important consideration than the gain in household output, but it depends upon the choice between income and leisure to be discussed in the next section.

Given that labour profiles differ for different crops, the work load for a combination of different crops is more level through the year than that for a sole crop. Thus another argument put forward in favour of mixed cropping is that it spreads the labour input more evenly over the year. In Northern Nigeria mixed cropping has been shown to yield a higher return per unit of peak labour input than sole cropping (Norman, 1974 op. cit.). However, the crops do not need to be grown in a mixture to achieve this benefit. Clearly, in bimodal rainfall areas with two distinct cropping seasons, the late maize crop does not compete with the early maize crop for either labour or land. The two crops are supplementary enterprises. Similarly, dry-season irrigation may be supplementary to rainfed cropping in the wet seasons, and may help to even out the work load, as shown in Figure 7.2.

The comparison of labour profiles, in which labour is treated as a fixed input per month and per hectare of a crop, is not the whole story. In practice, there may be scope for varying labour inputs on operations such as weeding or for delaying operations such as planting, although yields may be reduced thereby. Given that the labour profile represents the 'optimal' pattern and timing of the work for maximum yield, sub-optimal inputs on one peak-period activity may be justified in order to release labour for another. From Figure 7.1 we see that both clearing of new land and weeding take place in January and February on that particular farm. In effect, early planted land is ready for weeding before clearing and planting is

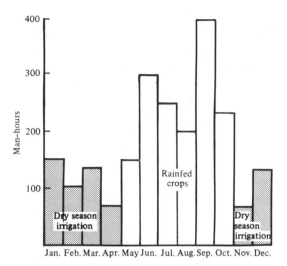

Fig. 7.2 Levelling labour needs with supplementary irrigation

finished. Time devoted to early weeding will raise yields, while time spent clearing new land will extend the area of crops grown. Thus, the farmer faces a nice problem in allocating labour and balancing the opportunity costs between these two peak-period activities.

Competition for labour at peak periods may arise between cash crops and food crops. One well documented example is cotton growing in both West and East Africa. Agricultural research in both regions has shown that substantial yield losses result from delayed planting of this crop. The fact that farmers do delay cotton planting is now recognized as reflecting their higher priority for the planting of food crops rather than ignorance or irrationality. New high-yielding cotton varieties, with a long growing season, are of little interest to such farmers. They prefer short-staple varieties which may be planted after the food crops without severe yield depression. This example demonstrates the importance of understanding the farmers' objectives and priorities, as well as identifying their critical constraints before embarking on crop breeding programmes, or indeed any agricultural research and development.

Another problem which farmers may face in the seasonal allocation of labour is the 'hungry gap' or pre-harvest shortage of food. This is most likely to arise in semi-arid areas where there is only one crop season, if insufficient food is produced to support the family over the whole year or if there are heavy losses in storage. The problem is particularly acute in that food energy intake falls and people are most susceptible to debilitating diseases at the busiest time of the year, when energy needs are greatest for crop work and for drawing and fetching water. If the labour supply at peak periods is reduced, opportunity costs and the problems of allocation are increased.

The farm work – leisure choice

Unlike the industrial worker, who is paid a fixed wage for a fixed number of hours of work, the farm family member is free to choose how much time he will devote to farm work and how much to leisure. Apart from the time needed for eating and sleeping, leisure is valued for its own sake; it is a form of consumption. Hence each hour of work has a subjective cost in terms of the leisure foregone. An individual will only work so long as he values the product of his effort more than he values the leisure he gives up.

In practice there is no clear distinction between work and leisure. Even where there is no off-farm paid employment, family members have many household tasks, such as fetching fuel and water, food processing, building or repairing dwellings and making various utensils, besides many religious and civic duties. All these goods and services are of value to the household and are sometimes grouped together as 'Z-goods'. Thus we now have a three-way classification of ways in which time may be spent; farm production, Z-good production and leisure proper. None the less, let us simplify by assuming that the choice is simply between farm work, which produces output on the one hand and leisure on the other. Let us also assume that the analysis relates to the allocation of time by a single individual over a particular period of the year. The conclusions may then be generalized.

Figure 7.3(a) illustrates the production response curve (OB) for varying labour inputs, at a peak work period, in maize production. Thus, it is similar to Figure 2.2. The corresponding marginal product curve is shown in Figure 7.3(b). Even though

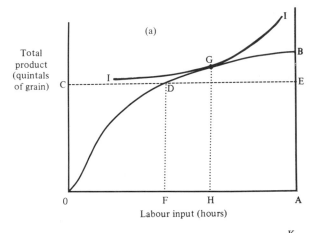

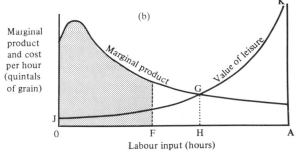

Fig. 7.3 The income–leisure choice, (a) Total product, (b) Marginal product

constant returns to scale may apply, it is appropriate to assume diminishing marginal returns to labour for a specific operation. The maximum number of hours that can be worked is set by the basic minimum leisure essential for eating and sleeping represented by the constraint line AB. Within the range OA increased work means less leisure. An additional constraint is set by the need to produce sufficient food for subsistence. This is represented by the line CDE in the upper diagram, and by the shaded area in the lower diagram. Thus, the feasible range of choice between work and leisure is limited to the range FA.

Utility is maximized at point G (in the upper diagram) where the indifference curve between maize output and leisure just touches the production response curve. (Note that the indifference curve slopes upward from left to right since a reduction in leisure is only acceptable if production is increased.

However, it may also be noted that the horizontal axis could be reversed to measure leisure rather than work, in which case the diagram would resemble the production possibility boundary of Figure 4.4.)

In the lower diagram, the line JGK represents the subjective marginal value of leisure, which increases as the hours of leisure diminish. The optimum utility maximizing choice occurs where the marginal value per hour of leisure is equal to the marginal value product of work, at G, when OH hours are worked.

Income and substitution effects

A rise in labour productivity as a result of new technology would have both a substitution effect and an income effect on the hours worked. The rise in the return per hour of work and the consequent rise in opportunity cost per hour of leisure provides an incentive to substitute work for leisure, by working longer hours. However, the income effect works in the opposite direction. With an increased income resulting from his increased productivity, a person is better off and can afford to 'buy' more leisure.

These effects are illustrated in Figure 7.4 in which the rise in productivity is represented by the upward shift of the response curve and the marginal product curve.

In the upper diagram, the new utility maximizing point M lies on a higher indifference curve, reflecting the rise in total income and utility. The effect on labour input in this example is a *fall* in the hours worked. This means that the income effect (PM) outweighs the substitution effect (GP), so the farmer decides to take more leisure.

From the lower diagram we see that if the subjective marginal value per hour of leisure remained unchanged, increased agricultural productivity would result in more hours worked (the substitution effect). However, the rise in income represented by the increased productivity causes a rise in the subjective valuation of leisure (the income effect), and this is sufficient to bring about a fall in hours worked. When a rise in productivity, or product prices, causes a reduction in hours worked, we have what is called a backward-bending labour supply curve. In the case of a product price increase, this may mean a decline in the product output, or negative price response.

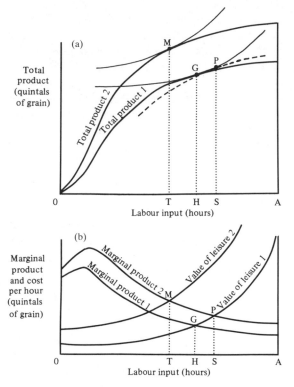

Fig. 7.4 Income and substitution effects, (a) Total product, (b) Marginal product

This analysis does not allow us to predict whether the income effect will be stronger and labour supply backward-bending, or not. There is no clear empirical evidence either, in relation to *total* labour supply. This is because, in practice, there is not just a simple choice between labour and leisure, but many different alternative activities. If the returns from one such activity increase, there will be a clear incentive to substitute this for other activities, while the income effect may be relatively small. Thus, if the price of one product rises there is generally a positive supply response, because labour is transferred from other activities.

Similarly, if we now compare the situation at seasonal work peaks and work troughs, the income effect is unimportant because income may be carried over from one period of the year to another. Thus, at work peaks the average and marginal product per hour for a specific labour input is much greater than at slack periods. This provides an incentive to work

harder and longer at work peaks and to take more leisure during slack periods. There may be scope for deferring some household tasks, like building and repairs, until the dry season when labour productivity on the farm is low. However, the numbers of hours worked on the farm on different days of the year are inter-related. For instance, the number of hours spent harvesting is dependent on the yield which, in turn, will be influenced by the time spent in cultivating and weeding.

The choice between work and leisure is likely to be influenced by the ease or difficulty of producing enough for subsistence. An individual cropping a small area of poor land and/or with a large number of dependants, is barely able to meet family food needs. He is forced to devote more time to farm work and take less leisure than his better-off neighbour. Following this argument, we might expect workers in a family with a high 'dependency ratio', that is a large proportion of dependent children and old people, to work harder than those with few dependants (see Chayanov, 1925). This could be represented in Figure 7.3(a) by raising the line CDE and flattening the indifference curve, so that the optimum would move nearer to point B. The number of hours worked would therefore increase. In Figure 7.3(b) the shaded area, representing basic food requirements, would represent a larger proportion of the total area under the marginal product curve, while the subjective value of leisure, relative to food, would fall, again resulting in a shift to the right of the optimum point. Evidence from many parts of Africa generally supports the theory that work input per person is directly related to the dependency ratio (see Norman, 1969, Levi & Havinden, 1982 op. cit. and Hunt, 1978).

Hired labour

Hired labour is of increasing importance with the spread of production for the market and increased individualism. The wages of hired labour make up the largest single item of expenditure on most farms, despite the fact that hired labour generally provides less than 20 per cent of the total farm work input. Since there are relatively few landless labourers in most parts of rural Africa, hired workers are generally farmers, either from the locality or migrants from

further afield. Migrant workers may live with the
employer's family and share his meals or they may be
allowed to establish food-crop farms of their own.
The so-called 'strange farmers' of Senegambia are
migrants, who have been granted land to grow cash
crops in exchange for work on the local farmers'
groundnut crop.

Share contracts may be viewed as a mechanism
whereby those who control the means of production,
land, tree crops or livestock, acquire access to the
labour of others. Their prevalence may reflect a
shortage of cash and a lack of short-term credit
facilities which would allow hire of labour for a
money wage. However, labour hire for wages often
co-exists with various sharing agreements in the same
village or area. There is no obvious tendency to
change from one form of contract to the other. As we
have noted earlier, share-cropping has certain advantages in spreading risks. From the farmer's point of
view it is most convenient to hire labour by the hour
as and when needed. However, the temporary piece-
worker hired by the hour has very little security and
no continuity of employment. Possibly for these
reasons, wages are generally highest for labourers
hired by the hour. This provides the simplest case for
analysis.

As explained in detail in Chapter 2, the profit
maximizing level of labour hire is found where the
marginal product is equal to the wage, measured in
terms of the product. However, when unpaid family
labour is also available, a choice must be made
between family and hired labour. The solution is
obtained by combining the analysis of labour hire
(Figure 2.3) with that of the farm work versus leisure
choice (Figure 7.3). The result is shown in Figure 7.5.
In the upper diagram (7.5(a)) the total labour cost
line (BCD) is a tangent to the production response
curve at point C. This then gives the optimum level of
labour input (OA hours) where the marginal product
is equal to the wage as described in Chapter 2. The
same labour cost line (BCD) is a tangent to the
farmer's indifference curve (II) at point Y. This
means that the farmer chooses to work only OX
hours, and hires the remaining XA hours.

The lower diagram (7.5(b)) shows the marginal
product curve, the horizontal line BD representing
the constant hourly wage rate, and the rising marginal

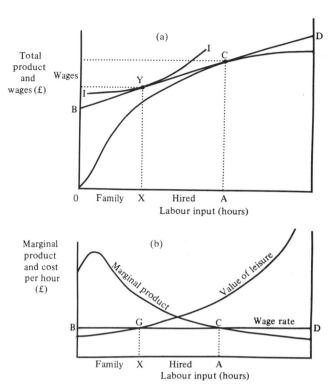

Fig. 7.5 Labour hiring, (a) Total product, (b) Marginal
product

value of leisure curve. Labour use is extended to
point C since the marginal product exceeds the wage
up to this point. Family labour is used up to point G,
since each hour of leisure is valued at less than the
wage over this range. The economic optimum of
family and hired labour is found when (a) the marginal value per hour of leisure is equal to the wage
rate and (b) they are both equal to the marginal
product of labour on the farm.

In fact, agricultural wage rates are themselves
influenced by the productivity of labour on farms as
well as the availability of labour. Where the marginal
product of labour is high, as in areas of high potential,
so too is the demand for labour. As a result wages are
likely to be higher than average in such areas. Im-
migration of workers, may increase the supply of
labour, and thus reduce the average wage, but this
process is unlikely to proceed so far as to eliminate
the wage differential. By a similar argument we might
expect wages of casual labour to rise at peak work

periods; as indeed often happens. It may also help to explain why women and children are paid lower wages than adult males for certain tasks, where their productivity may be lower.

Another consideration is that some tasks, such as bush clearing, are more arduous than others and a higher wage, which includes a 'compensating differential', is necessary to attract sufficient labour. Thus quite large wage differences may exist between regions, sexes and seasons of the year. (See Byerlee *et al.*, 1976). Other things being equal, a rise in wages would induce a reduction in the desire to hire labour, although the effect on family labour input is uncertain. However, if wages are high enough, farm family members may be persuaded to seek off-farm work.

Off-farm work

Figure 7.5 is readily modified to show the allocation of family labour to off-farm work as in Figure 7.6. The only difference is that the wage rate is higher, so that the optimum level of family work is greater than the profit maximizing level of labour input on the farm (point Y is now to the right of point C). Now OA hours are worked on the farm and AX hours off the farm. The same effect would result if the production response curve for farm labour was lower than in Figure 7.5, or the utility curve was flatter, reflecting a higher preference for income rather than leisure. If the off-farm wage were higher than the marginal product of labour in farming at all levels of employment, the best policy would be to leave farming altogether.

Most farm families in Africa are involved in off-farm work or at least some household members are. Off-farm activities represent an alternative form of employment and source of income, which must also be taken into account in considering the opportunity costs of different farm enterprises.

Since the marginal product or opportunity cost of farm labour varies over the year, so do the relative attractions of off-farm work. Ideally, off-farm work would be fitted in during slack periods, while labour would be hired-in for work peaks. Thus hiring-in *and* hiring-out of labour may occur on the same farm at different periods of the year. In fact, the wage rate for off-farm work may differ from that for hiring

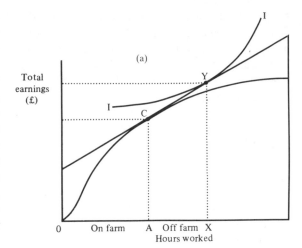

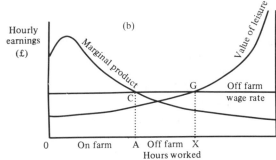

Fig. 7.6 Off-farm work, (a) Total product, (b) Marginal product

agricultural labourers, in which case it may pay a family to hire farm labour at the same time as some family members work elsewhere. Three main categories of off-farm work may be identified:

(i) local employment on other farms in local crafts such as blacksmithing or other occupations such as trading;

(ii) urban employment in a neighbouring town which is of particular importance to those dwelling on the urban fringes and people like the *Yoruba* of Nigeria who have traditionally lived in towns and farmed the surrounding hinterland;

(iii) long-range migration of some family members to work in cash-cropped areas, on plantations or in the mines.

In the latter case the migrant faces the risk of not finding employment when he has made the move.

Thus he really needs to estimate his 'expected wage' (the going wage rate in the region to which he is migrating multiplied by the probability of finding employment) to compare with the marginal product in agriculture (which is also an expected value) in assessing whether migration is justified. Due allowance must also be made for travel costs, of course.

Most part-time farmers are loth to give up farming altogether. Their land holdings represent links with their families and communities and provide security as a basis for subsistence if the off-farm employment should cease. However, they may devote less time and effort to the land, than their full-time neighbours. Because of differences between the sexes and age groups in off-farm work opportunities, those most likely to move out of agriculture, if only temporarily, are the young adult males. As a result, women face increased responsibility for farm work, and increased work loads, while the dependency ratio on the farm may rise. Remittances of off-farm earnings may resolve the latter problem, but the absence of adult males creates shortages of labour for crucial tasks such as land clearing. As a result land already cleared, is cropped for more years than normal with resultant declining yields and loss of soil fertility (see Hunt, 1984, Swindell, 1985).

Increasing labour productivity

Labour productivity, in terms of the average product per person employed in agriculture, may be increased either by producing more with the existing work force or by saving labour. This distinction is somewhat artificial since labour which is saved might be used to increase output without increasing the work force. However, this is only likely to occur if labour is saved at peak work periods, when seasonal availability is an effective constraint. Labour saved at slack periods simply adds to leisure time. The real distinction to be drawn is that between neutral innovations, such as some new high yielding crop varieties, which make a net contribution by raising the productivity of all inputs equally and labour-saving innovations such as mechanization which generally involve the substitution of capital for labour (see Chapters 2, 3, and 6 for analysis of technological change).

Labour-saving innovations may include herbicides, zero-tillage techniques, animal draught and various levels of mechanization. The introduction of any of these necessitates capital investments; special equipment is needed for applying herbicides or for seeding through the herbage cover under zero tillage, while oxen, ploughs and machines are also items of capital. The introduction of ox ploughs and tractor mechanization also involves the initial costs of training oxen and operators and clearing and destumping the land. This last cost item may be prohibitive in humid rain-forest areas.

There is no assurance that the benefits of such labour-saving innovations will outweigh the costs. First, the elimination of one work peak may simply result in another becoming critical. Thus the introduction of zero tillage, animal draft power or tractors may reduce or eliminate the work peak for cultivation and planting, only to leave weeding labour as a critical constraint. The use of herbicides reduces labour needs for weeding but then harvest labour may become critical. As a result there may be very little saving in the labour force required. Second, even when a real reduction in labour needs is achieved there is no benefit, other than the reduction in drudgery, unless the labour saved can be used productively. There are three possible ways of absorbing labour which is saved: i) intensifying and increasing yields per hectare, ii) extending the cultivated area, and iii) finding alternative off-farm work.

In principle, the use of labour-saving technology, especially machinery, should allow greater cropping intensity and increased yields as a result of more timely and effective crop operations. There is relatively little evidence of such benefits being obtained in practice. It is suggested that the introduction of tractors on the Mwea Irrigation Scheme in Kenya was accompanied by increased rice yields. None the less in such cases it is not easy to separate the effect of the labour-saving technology from other changes such as improved management and increased use of fertilizers. It must be concluded that, even if there is scope for increasing yields by the use of machinery, it is limited.

The alternative of extending the cultivated area may be feasible if there is an abundance of unused land, although since ox-teams and tractors are lumpy, indivisible items, a substantial increase in farm area may be needed to cover the costs (see discussion of farm scale in Chapter 3). Where this entails a lengthening of the cropping period and a shortening of fallows, yields are likely to diminish unless alternative means are found of restoring fertility. Indeed, as shown in the last chapter, this represents intensification of land use rather than an extension of the area used. This in turn, of course, implies that land is now a limiting constraint, but it does not necessarily conflict with the argument that peak labour is the *most* limiting constraint. It is usually the case that if the most limiting constraint is overcome, another secondary one becomes effective. Land is highly likely to become an effective constraint given the large increase in scale associated with mechanization.

The expansion of off-farm employment makes no direct contribution to agricultural output of course, but it may be beneficial in raising rural incomes. Opportunities for part-time farming may allow some families to remain in agriculture, who otherwise could not survive. Thus the promotion of improved transport to urban areas or of rural and cottage industry may help to raise rural welfare. Finally the provision of improved services, notably water supplies, rural electricity and health centres may improve farm labour supply throughout the year. Apart from the general gain in welfare, farm production may be increased as a result of the labour released for farm work at peak periods.

In summary it is clear that labour productivity may be increased through the introduction of new technology but additional costs are always incurred. Careful economic evaluation is needed before any innovation is recommended for general adoption. Farming Systems Research and the planning methods discussed later in this book may be used to evaluate specific proposals.

Further reading

Byerlee, D., Eicher, C. K., Liedholm, C. & Spencer, D. S. C. (1976). *Rural Employment in Tropical Africa: Summary of Findings*, Michigan State University, Department of Agricultural Economics, African Rural Economy Working Paper 20

Chambers, R., Longhurst, R., Bradley, D. & Feachem, R. (1979). *Seasonal Dimensions to Rural Poverty: Analysis and Practical Implications,* Sussex, England: Institute of Development Studies. Discussion Paper 142

Chayanov, A. V. (1925). *Peasant farm organization* in *The Theory of Peasant Economy*. Eds Thorner, D., Kerblay, B. & Smith, R. E. F. (1966), Homewood, Illinois, Irwin

Cleave, J. H. (1974). *African Farmers: Labour Use in the Development of Smallholder Agriculture*, New York, Praeger

Farrington, J. (1975). *Farm Surveys in Malawi: The Collection and Analysis of Labour Data,* University of Reading, Department of Agricultural Economics: Development Study 16

Hunt, D. (1978). Chayanov's model of peasant household resource allocation and its relevance to Mbere Division, Eastern Kenya, *Journal of Development Studies,* **15**(1)

Hunt, D. (1984). *The Labour Aspects of Shifting Cultivation in African Agriculture*, Rome, FAO

Hymer, S. & Resnick, S. (1969). A model of an agrarian economy with non-agricultural activities, *American Economic Review,* **59**(4) 493–506

Lagemann, J., Flinn, J. C., Okigbo, B. N. & Moormann, F. R. (1975). *Root Crop/Oil Palm Farming Systems: A Case Study from Eastern Nigeria*, IITA, Ibadan

Low, A. R. C. (1986). *Agricultural Development in Southern Africa: Farm–Household Economics and the Food Crisis*, London, James Currey

Nakajima, C. (1986). *Subjective equilibrium theory of the farm household*, translated by R. Kada. Amsterdam, Elsevier

Norman, D. W. (1969). Labour inputs of farmers: a case study of the Zaria Province of the North Central State of Nigeria, *Nigerian Journal of Economic Social Studies* (II) 3–14

Swindell, K. (1985). *African Society Today: Farm Labour*, Cambridge University Press

8

Capital and credit

Types of capital

Everything used in production, which is not a gift of nature but has been produced in the past, is called capital. It includes not only machines and tools but also buildings, roads, footpaths, drainage ditches, terraces, irrigation equipment, growing crops, livestock and stocks of food, seed, fertilizers and other materials. Clearly, many very different items are included. All they have in common is that they were produced in the past and will contribute to production in the future.

It should be clear that some capital is needed for any kind of productive activity. For instance, the spears and food and water containers of pre-agricultural, food-gathering societies are items of capital. A typical arable farmer may own three cutlasses or machetes, two hoes, an axe and a grain store which may together be valued at about £20. Capital valuations are often much higher where permanent crops are grown, livestock are kept and machinery used. Each item of capital controlled is known as an asset.

We are concerned here with the assets used in the process of agricultural production but two other forms of capital should be noted in passing. One is social overhead capital, which includes communications, market-places, public utilities, research stations and agricultural extension services, and is best considered as a feature of the farmer's environment. The other is consumer capital, made up of durable consumer goods such as houses and furniture. It may be difficult to decide whether a particular asset is productive or consumer capital. For example, a bicycle may be used for pleasure or for transporting farm produce to market. A house, furniture and cooking utensils are necessities which must be available before man is capable of productive work. The problems of definition are acute in family farming where there are close links between farm and household.

A farmer's capital assets may be classified according to the length of their productive lives into long-, medium-, or short-term capital. Long-term capital has a life of many years and may be virtually permanent. It includes items like buildings, wells, dams and land improvements. Certain tree crops may come into this category. Capital with a medium-life span of just a few years includes workstock such as bullocks, breeding and milking stock and many items of tools and equipment. Short-term capital is generally consumed within one year and includes stocks of food, seed, agricultural chemicals and cash. The harvesting of annual crops usually occurs in one or two discrete periods of the year so not only do stocks of seed have to be provided some months before benefits are obtained but also family consumption must be met during the period between one harvest and the next. For these purposes capital in the form of stored seed and food (or the cash to buy them) is needed for at least part of the year. This short-term capital is also known as circulating or working capital to distinguish it from other assets which are not consumed within a single year and are therefore known as fixed or durable capital.

Capital is made up of so many different items that it can be misleading to treat it as a single resource. Land and labour vary in quality, as we have seen, but it is generally reasonable to assume that these resources can be transferred from one enterprise to another, say from cattle grazing to cotton production, without too much difficulty. With capital it is less straightforward. Since capital is embodied in durable physical assets, transfer from one use to another is difficult. They are fixed items of cost.

Investment and saving

Investment means adding to the stock of capital. It can take place in several apparently different ways, although in effect they all amount to the same thing, namely, saving, which means foregoing current consumption. The first way in which a farmer invests is by actually saving some of his produce. For instance, cereals, legumes or yams which are stored either for seed or future consumption represent an addition to the stock of capital, and are therefore investments. Goats and cattle which are kept for milk or breeding or just to fatten into bigger animals also represent savings and investment. Secondly, assets are created by the farmer's own physical efforts. If he clears land, plants trees, builds a dam or a cattle kraal he is investing. Again he foregoes current consumption because he might have spent his time either in producing more food or in leisure, which in itself is a form of consumption. Investment by foregoing leisure is particularly suited to the small farmer. Especially if his total output is little above the subsistence minimum he cannot afford to forego consumption of produce, but even the poorest producer will almost certainly have some leisure time, particularly during the seasonal troughs in the pattern of labour requirements. The opportunity cost of this labour may be very small indeed.

The third method of investment is by purchase. This can, of course, only occur in an exchange economy where produce is sold or bartered, but this is true of most of Africa today. Again, current consumption is foregone if the money or bartered goods would otherwise have been used for consumption purposes. Certain assets cannot be manufactured on the farm and must be purchased. This is particularly true of tools, machinery, stocks of improved seed and agricultural chemicals. It may occur to the reader at this point that capital assets are sometimes hired or purchased with the aid of a loan. We will return to these possibilities later but it is worth noting that hired or borrowed capital is still the outcome of saving by someone other than the user. Furthermore,

although investment is ultimately dependent upon saving it does not follow that all savings are necessarily invested. They may be used for consumption, for festivities or hoarded as a reserve against risk.

Saving, no matter for what purpose, represents a cost to the user, namely, the cost of waiting. Why then do individuals save? The answer is that the benefits that are obtained by waiting are greater than the value of the consumption foregone. In the case of productive investment, greater returns are obtained in the long run by using indirect or roundabout methods of production. Thus the farmer who spends time making a plough, training bullocks to draw it and destumping his land must work harder than his neighbours who use hoe cultivation, and he may produce less food than they while he is making this investment. In future years, however, he hopes that plough cultivation will add to his output more than enough to make up for his original efforts. This surplus, over and above the cost of the investment, is known as the 'return on capital'.

Return on capital from goats

Most rural families and some urban dwellers in much of Africa keep a few sheep and goats. These small ruminants are often left to forage for themselves in free-roaming village flocks with minimal management. Labour and other costs are very low, but mortality may be quite high. Sheep and goats kept for breeding are clearly capital assets. They have been produced in the past, and will contribute to future output by producing young. Thus we may illustrate the concept of return on capital by considering production from a small goat flock (see Upton, 1985).

West African dwarf goats, of the rain-forest region, are highly prolific. They reproduce every eight or nine months on average, frequently bearing twins. As a result each doe bears two or more kids each year. However, mortalities are such that only half the kids survive to maturity at about 12 months of age. Thus a typical household flock of four does would produce four surviving animals each year. Predicted flock structures at yearly intervals over two years if there were no off-take, are presented in Table 8.1. In practice it is not necessary for every family to keep a buck since each one may serve about 20 does. An

Table 8.1. *Goat flock structure over two years with no offtake*

	Initial flock	End of one year	End of two years
Adult female	4	4	6
Adult male	1	1	3
Immature female	—	2	2
Immature male	—	2	2
Total	5	9	13

adult male is included in this simplified example to avoid dealing with fractions of an animal. Note that we are arbitrarily dividing continuous time into yearly intervals. More accurate results could be obtained by dividing time up into shorter intervals and correspondingly increasing the number of age cohorts. However, this would necessarily lead to increased occurrence of fractional numbers in the calculations.

The number after one and two years are predicted by assuming each doe bears one surviving kid annually and that half of all kids born are female. By the end of the second year the kids born in the first year have matured and joined the stock of adult goats; while more immatures have been born. The mortality of adult goats is ignored for the present.

Clearly, with this set of assumptions the flock would grow steadily over time. Alternatively, from the end of year one, the family could consume four animals, two males and two females, each year and maintain a constant flock size of nine animals. On this basis a capital stock of five breeding animals appears to yield a return of four animals per year.

However, this could not continue indefinitely since eventually the breeding stock will need replacing. Given that, on average, adults spend five years in the breeding flock, this is the maximum period that an offtake of four animals per year could be maintained. Allowances for replacement are necessary if a steady-state, constant-sized flock is to be maintained. Of course the productive lives of individual animals vary considerably around the average of five years. Thus it is more convenient to think of the replacement cost as an annual rate of 20 per cent. There is then an annual cost of replacing one-fifth of the adult animals, which in our example means one replacement every year.

This annual replacement cost is generally known as 'depreciation'.

We now adjust the results for depreciation of breeding stock, to give an annual offtake of three animals (four minus one) from a steady-state breeding flock of six animals (five plus one replacement). The annual rate of return on capital is therefore

$$\tfrac{3}{6} = 50 \text{ per cent}$$

This is a rather crude estimate based on rounding-off the productivity estimates and assuming all animals are equally valued. In fact, the value of a one-year old animal is less than that of a breeding adult. Thus the rate of return in money value terms is lower; estimated at 34 per cent by Upton (1985). However, for present purposes the cruder estimate based on animal numbers may be used.

Note that we now have two alternative ways of viewing a capital asset, either as a stock of past production, in our example six goats already produced, or as a stream of future incomes, in this case three kids per year continuing indefinitely into the future. This case represents a very simple example of a 'flock growth model'. To build such a model, the flock structure is described by the numbers (proportions) in each of the various age and sex cohorts. The new flock structure after a given time period has elapsed is predicted on the basis of productivity and mortality parameters. Thus the number of kids is determined as the number of breeding does multiplied by the kidding rate and multiplied by the survival rate for kids. The number in an older age group is given by the number in the next younger group in the previous period multiplied by the survival rate, and adjusted for off-take if any, (see Figure 8.1.) In this way the development of the flock structure over time may be predicted for different off-take policies or different production parameters (Upton, 1985).

Intertemporal choice

In this example the steady-state policy was emphasized. A depreciation allowance was made to allow for the maintenance of a constant breeding stock. Net investment (ie net of depreciation) or capital formation, was assumed to be zero. This is a convenient approach for estimating the rate of return

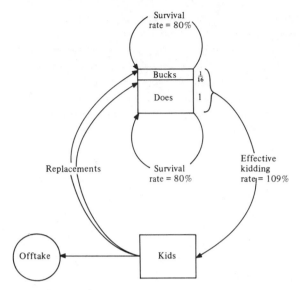

Fig. 8.1 Goat flock growth model

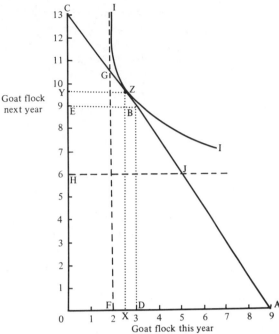

Goat flock
next year

Goat flock this year

Fig. 8.2 The choice between present and future consumption

but maintenance of a constant breeding flock is not the only feasible policy. Other feasible alternatives are shown in Figure 8.2. The farm family could, for instance, have a grand feast and consume all nine goats in the first year. They would then have no goats left in year two (point A). The steady-state alternative of consuming only three animals and retaining the rest for breeding purposes is represented by point B. Another alternative would be to consume none and retain all the offspring to maximize flock growth (point C). If we assume that numbers of animals can be varied continually (which may be justified if treated as expected values rather than actual numbers) intermediate policies may be represented by other points on the line ABC. This may be viewed as a production possibilities boundary for the two commodities, goats this year and goats next year.

Along the horizontal scale, the distance OA represents the flock or stock of capital in year one. The amount consumed is measured along this scale from the origin (OX), the remainder (XA) being retained or invested for future production. When XA units are retained, the stock next year will be OY goats. Clearly, there is a trade-off between off-take of goats this year and the number of goats available next year, so the boundary has a negative slope. The (negative) slope, or RPT, for this boundary is equal to (1 + r)

where r is the rate of return. Thus between points A and B the (negative) slope is

$$\tfrac{9}{6} = 1.5. \qquad \text{Hence } r = 0.5 \text{ or 50 per cent}$$

The rate of return between B and C is lower since three more goats invested produce only four additions to next year's stock. The slope is $\tfrac{4}{3} = 1.33$ so r = 33 per cent. This decline in the rate of return reflects diminishing marginal returns to increased investment. (The perceptive reader may question how thirteen goats can be produced if one is needed for replacement. The answer is that the additions to the flock are assumed to start breeding from twelve months of age.)

Intertemporal choice may be illustrated on this diagram along similar lines to our earlier analyses (particularly Chapter 4). First, the family may have a target consumption need of say two goats per year represented by the vertical line FG. This restricts the amount of investment and hence the size of next year's flock to about ten animals. Note that, if more than three goats are consumed, disinvestment occurs and the breeding flock is reduced.

The family may also aim to retain the flock of at least six animals, represented by the constraint line HJ. Given the other constraint FG, the choice is limited to the segment of the boundary GJ. If we assume that the farmer's utility depends upon his expected future consumption as well as his current consumption, indifference curves may be drawn, showing the subjective trade off between goats this year and goats next year as shown. The (negative) slope is equal to $(1 + p)$ where p is the rate of subjective time preference also known as the 'personal discount rate'. When p is positive it implies that current consumption is valued more highly than future consumption.

The optimum choice is again found at the point of tangency Z. This represents the consumption of OX goats ($2\frac{1}{2}$ on average) and investment XA ($6\frac{1}{2}$ on average) this year to produce a flock of OY (nearly ten goats) next year.

At the optimum point, the slopes of the indifference curve and the production possibility boundary are equal so the rate of time preference equals the rate of return. This rate of return, estimated here at 33 per cent, therefore represents the opportunity cost of capital. Any other investment which is expected to earn a return of 33 per cent, or more, should be adopted. Ultimately, an optimal allocation of scarce capital is achieved when marginal return, that is the rate of return on the last unit of investment, is the same for every use, both on and off the farm.

It should be noted that the choice between consumption and investment also determines the growth rate. From Figure 8.2 it is apparent that when consumption is OX so that net investment is XD, the growth rate is EY/OA, which is about 8 per cent. Similar decisions for other investments, in aggregate, determine the overall growth rate of the family income.

Indivisibilities, risk and circulating capital

These statements regarding the allocation of capital really need qualifying to allow for indivisibilities and risk. Even goats are indivisible, and it will be noted that in our example the optimum allocation involves consuming two and a half goats! This might possibly be achieved by consuming two animals in one year

and three the next, or vice versa. However, it is clearly not possible to attain exact equality of marginal returns for all uses in every year. The opportunity cost principle may still be applied; namely, to maximize utility, each unit of capital should be used where it will earn the highest return.

There is a further problem in that capital already invested in goats cannot readily be transformed into another form of capital which might earn a higher rate of return. Once capital has been invested in durable assets it is committed; flexibility is lost. Only when new investments are planned is there scope for adjusting the allocation to maximize utility.

However, when new investments are planned, the outcome is uncertain; all such decisions are necessarily subject to risk. Although a farmer may allocate his investments so as to equalize *expected* marginal returns in all uses, he cannot be certain that the *actual* marginal returns will be equal. Indeed, if he is risk averse he will not base his decisions on the expected returns only. Generally, the riskier a project, or the greater the potential variation in outcomes, the higher must be the rate of return to justify the investment (see Chapter 5). A crude but simple method of allowing for risk in this context is to arrive at a 'certainty equivalent' by means of a 'risk-discount'. For example, suppose the expected rate of return is 50 per cent, then it may be reduced by a risk discount of ten points to a figure of 40 per cent for a mildly risky investment, or by a risk discount of 20 points to give a 'risk-discounted return' of 30 per cent for a more speculative project. The risk-discounted value of the prospective return is then the basis for allocating new investment. A low expected-return, low-risk investment might be preferred over a potentially more profitable but riskier choice.

Returns to working capital cannot be calculated in the same way as outlined above, because it is used to provide other inputs such as labour. Thus it is impossible to distinguish the marginal return on working capital from the marginal product of the labour. None the less, working capital is essential for continued production, so in that sense it is productive.

Requirements of working capital vary within each year according to a fairly regular cycle. Over the cropping season the costs of seed and of supporting the labour force must be met some weeks or months before the returns are obtained at harvest time. Thus the working capital requirement usually rises to a peak just before harvest. If the produce is stored after harvest the capital requirements continue to rise until the produce is finally sold or consumed. However, the peak capital requirements of different enterprises may occur at different times of the year. Where this is the case, the working capital requirement for a combination of enterprises will be more level through the year than the requirement for a single crop. Enterprises may be supplementary in the use of working capital in much the same way as they may be supplementary in the use of labour.

Stored produce may earn a return in the sense that it gains in value as the dry season progresses and food becomes more scarce. Against this benefit must be set the costs of storage, which are the capital costs of the storage barn or granary and the storage losses due to insect, fungal and rodent damage. Grain storage losses in the drier Savannah regions may be as low as 3–4 per cent, but they are higher in more humid zones and for cowpeas and root crops. None the less, storage can generally be justified economically since the gain in value is sufficient to cover costs and earn an acceptable return on the capital invested in the store and the produce itself, (see Upton, 1962). Note that in this case the return is earned over a part of a year and needs adjustment to give the equivalent annual rate.

Discounting for longer-term investments

For longer-term investments an alternative method of evaluation is needed. To illustrate let us consider a rather artificial example of a calf bought purely for fattening and sale three years later. This animal is supposed to forage for itself so there are no feed or labour costs as was assumed for the goat flock. However, the value of the animal is now measured in money terms.

The use of money values was avoided in the goat example in order to emphasize the physical productivity of capital and the idea that even pure subsistence farmers face capital investment decisions. However, in the new example the benefit is simply a gain in value. Energy value of the carcass might have been used, where the animal was kept for home

Table 8.2. *Compounding investment costs*

	Cumulative cost	Interest	Total
Outset	20	6	26
End year 1	26	7.80	33.80
End year 2	33.80	10.14	43.94
End year 3	43.94		

consumption, but we assume the calf is purchased for cash and later sold. The purchase price of the calf is £20 and the expected sale price three years later is £60. Clearly there is a gain of £40 over the three years but it is not yet clear whether this is sufficient to justify the investment. The opportunity cost of investment funds is 30 per cent or 0.3.

The results in Table 8.2 show how the cumulative cost of the initial investment and its opportunity costs increase over the life of the animal. After the animal has been kept for one year the cumulative cost C, of purchase (V = £20) and the opportunity cost (pV = 0.3 × £20) is £26 given by

$$C_1 = V(1 + p)$$

In the following year the opportunity cost applies to this total sum so that total cost after two years C_2 is

$$C_2 = C_1(1 + p) = V(1 + p)^2$$

Similarly after three years

$$C_3 = C_2(1 + p) = V(1 + p)^3$$

This process of predicting accumulated future costs over time is known as 'compounding' and is the method used for adding 'compound interest' on bank loans. The general formula for an initial sum V, and an opportunity cost rate p calculated over n years is $V(1 + p)^n$. Thus £20 compounded over three years at 30 per cent amounts to

$$£20(1 + 0.3)^3 = £43.94.$$

On this basis this project is clearly worth undertaking. The cumulative cost at the end of three years is £43.94, although it sells for £60. Thus there is a terminal net gain of £16.06 (= £60 − 43.94). However, this gain will not be obtained for three years. For the farmer deciding whether to make the invest-

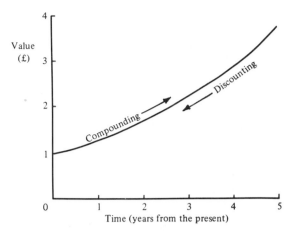

Fig. 8.3 Discounting: the converse of compounding

An investment which yields a positive terminal net gain must also yield a positive net present value (NPV). Thus on either criterion it is worth undertaking, provided that the associated risk is acceptable.

We now have a method of defining the value of any capital asset. It is the discounted, present value of expected net benefits. Thus the present value of this beef animal after keeping it for one year is $60/(1.3)^2 = £35.5$. The discounted present value of a plot of land expected to produce an annual net return of £90 per year is obtained simply by dividing by the discount rate thus

$$£90/0.3 = £300$$

In short £300 invested at 30 per cent would earn £90 per year.

The internal rate of return and replacement

In this calf-fattening example there is clearly a 200 per cent return over the three years; £40 gain in value for a £20 investment. The annual rate of return r is easily calculated by discounting so that

$$\frac{60}{(1 + r)^3} = 20 \text{ or compounding } 20(1 + r)^3 = 60$$

Which gives $r = 0.44$ or 44 per cent. This is known as the internal rate of return (IRR) and is that discount rate which results in a net present value of zero.

The internal rate of return can only be calculated in this way when there is a single input cost followed, after some years, by a single lump sum receipt as in our example. Generally it is necessary to use a trial and error procedure, calculating the NPV for different discount rates and interpolating as shown in Figure 8.4.

Real-world investments generally involve a series of costs, (eg annual food costs for the fattening calf) and may yield a series of benefits (eg milk from a cow or draught power from an ox). All these costs and benefits need to be evaluated in the same terms, usually money, and the resultant cost and benefit streams must be discounted to find the NPV or the IRR. Details of this planning procedure are given in Chapter 15. For present purposes, it must be recognized that a farmer faces difficult judgements in making long-term investment decisions. In so far as

ment it would be more useful to know what the gain is worth now; what is its 'net present value'. This is easily arrived at by 'discounting' which is simply the converse of compounding (see Figure 8.3). Thus if £1 now is worth £1.30 next year (if compounded at 30 per cent) then £1.30 received next year is only worth £1 now (if discounted at the same rate). In short since investment has an opportunity cost, £1 in the hand now is worth more than £1 expected in the future. Similarly, given that the future value F of a sum V compounded at a rate p over n years is

$$F = V(1 + p)^n$$

by rearranging, we find the present value of F is V thus

$$V = \frac{F}{(1 + p)^n}$$

Using this formula or the discount factors from Appendix Table I we find the present value of £16.06 in three years' time at a discount rate of 30 per cent to be

$$\frac{£16.06}{(1.3)^3} = £16.06 \times 0.4552 = £7.31$$

In fact, this could have been estimated more simply by discounting the £60 expected revenue, to give a present value of £27.31 ($=£60 \times 0.4552$) and subtracting the initial investment of £20.

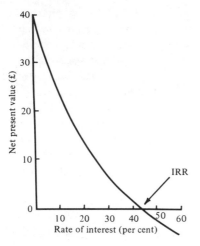

Fig. 8.4 Finding the internal rate of return

he can 'judge' the IRR, he can use it as described earlier to arrive at the optimum allocation of new investment.

A further capital investment decision problem may be illustrated using our simple example; that is to find the optimum age for replacement. How long should the investment last? At first sight it might appear that the investment should continue so long as further gains can be made. However, this is not the case since investment funds have an opportunity cost. The general rule is to adjust the life of the investment so as to maximize the annual gain.

Let us suppose our beef animal could be kept for a fourth year at the end of which it could be sold for £85. Thus it continues to gain in value by £25 (£85 − £60) in the fourth year. However, this gain represents only a 42 per cent return on the £60 value of the animal. Since the opportunity cost of investment funds is 44 per cent, which could be earned by investing in another calf and keeping it for three years, there is no justification for keeping the original animal for a fourth year. Thus we are assuming that 44 per cent is the maximum annual return, or opportunity cost of funds.

The cost of saving and the supply of capital

So far we have been considering the allocation of a fixed stock of capital but a farmer can, of course,

increase his capital resources by saving. As we have seen, saving involves the cost of waiting; most individuals would prefer to receive a quantity of a particular commodity, say ten bags of rice, now rather than the promise of the same amount a year from now. However, individuals vary in their time-preference; to be persuaded to wait a year one man may require the certainty of getting 11 bags of rice, while another would want 12. This means that the first man wants to be sure of getting 1 extra bag in 10, that is a 10 per cent return. The second man wants a 20 per cent return. These attitudes would be reflected in the slope of the indifference curves shown in Figure 8.2. Thus this personal time preference sets a lower limit on the risk-discounted return a farmer will accept in deciding whether to invest. Naturally, the higher a man values his present consumption in relation to future consumption, the higher this minimum acceptable rate of risk-discounted return will be.

It should not necessarily be inferred from this that if, because of improved technology or improved product prices, the risk-discounted return on capital rises, farmers will save more. The supply curve for savings may be backward sloping as is possible for labour. A rise in risk-discounted return increases the income of the investor and he may consequently prefer to increase his level of consumption and therefore save less. Indeed, the level of risk-discounted return probably has relatively little influence on the amount saved from any given level of income but much influence on the use of savings. The amount saved is likely to be affected much more by the level of farm incomes and the saving habits of the farming population. The saving habits of a person or a community are measured by the 'marginal propensity to save' which is the proportion that is saved from each additional £1 (or other unit) of income.

It is usually found that rich people save more than poor people, not only in absolute amounts but also as a proportion of their total income. The very poor are unable to save at all. Instead, they 'dis-save' or spend more than they earn, the difference being covered by going into debt or using up previously accumulated savings. As incomes rise, so too does the marginal propensity to save. Thus, if we compare African farmers with wealthier societies elsewhere, we find that saving and capital investment per person is low because incomes are low, but it may be argued that incomes are low in turn because the amount of capital per person is low. This is known as 'the vicious circle of poverty' because it implies that poor people must remain poor unless capital is introduced from outside to break the circle. It also implies that poor farmers will have a greater preference for present consumption than will wealthier people. The poor farmers are more concerned with survival from day to day until the next harvest rather than with investment for the future. This means that they will only invest in activities which are expected to yield a relatively high rate of risk-discounted return. All we are saying really is that where capital is scarce in relation to labour and land, it will be costly in relation to these other resources.

Other reasons for saving and investing are i) to create a reserve against risk and ii) to help children to become established on their own and to provide for old age. Some fruit orchards and poultry units are established to provide a low labour-input source of income after the owner is too old for full-time work.

So long as there is some net investment every year the total stock of capital will increase continually. This means that if there are diminishing returns to extra units of capital used with a fixed area of land and a fixed labour force, the rate of risk-discounted returns on the additional investments will fall over time. Eventually, the rate of return may fall to a minimum acceptable level where no further investment is justified. Thus a point of stagnation might be reached where farmers are capable of acquiring more capital by saving but where there are no further opportunities for productive investment.

In practice, the situation is complicated by the process of change. As shown in Chapter 4 some net investment is necessary to keep pace with family growth even if there are no innovations; but where agricultural change is taking place, new investment opportunities are constantly being introduced. Such technical innovations generally yield higher risk-discounted returns per unit of capital than do extra investments in traditional productive activities. Thus technical innovations provide new opportunities for productive investment, to which some farmers may respond without outside assistance.

A good example is the rapid expansion of cocoa production in West Africa during the last half-century. The massive investment in establishing the trees was provided by the farmers themselves as they became aware of the possibility of growing this new crop for export. Similarly in Kenya more recently the establishment of tea and coffee was largely financed by the farmers themselves. In fact, among cash-cropping farmers the rate of saving and investment is often high by any standards. Studies in Nigeria, Zambia and elsewhere have suggested that they may save as much as one-third of their incomes.

Borrowing and lending

Opportunities for borrowing and lending may increase farm incomes and welfare as shown in Figure 8.5. This is based on Figure 8.2, but now the line CK has been added to represent the possibility of borrowing or 'caretaking' of breeding goats. This practice is very common in West Africa, where the payment for the loan consists of half the offspring which are returned to the owner. Since we have estimated a

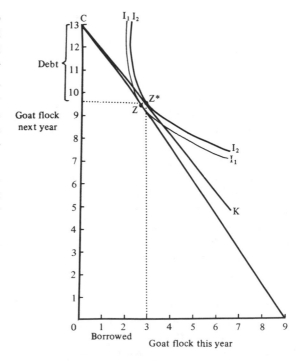

Fig. 8.5 Gains from borrowing

return of 50 per cent from goat breeding, the hire charge is estimated at half this, namely 25 per cent. The hire charge represented in this way is similar to 'interest' charges on a financial loan.

Given this alternative prospect, the optimal policy is to maximize investment, and keep nine goats for breeding (point C) but to maintain consumption levels by borrowing two or three animals each year. In this way the household utility is raised from point Z to Z*. Next year the flock will have increased to 13 animals but three of these are owed to the lender. Note that in this case, point C is a corner solution, where the rate of return exceeds the interest rate. If the farmer could increase his herd to more than nine animals, by borrowing more, it would pay him to do so.

The question may be asked why the owner is happy to lend animals on terms which are rather unfavourable to him. There may be an element of social duty to poorer kinsmen, but there may also be physical constraints limiting the size of flock an individual can conveniently keep. He is then happy to make *some* return by lending.

However, in some circumstances utility may be increased by lending as shown in Figure 8.6. Now the production possibility curve represents money values of wealth this year and wealth next year, and is drawn as a curve to illustrate diminishing marginal returns to investment. In this case the line FK represents the lending activity. Its (negative) slope is (1 + i) where i is the rate of interest.

The optimal solution now is for the farm family to consume quantity OX, to lend quantity XD and invest quantity DA. Next year total assets will amount to OE produced on the farm plus EY, the loan with interest. This choice of activities ensures that the slopes of the production possibility boundary, the lending activity line and the indifference curve are all equal. Thus the marginal rate of return on investment, the rate of interest and personal time-preference rate are all equal. The present value of the entire system, discounted at this rate, is given by OK. Similar analysis would apply to off-farm investments offering a constant rate of return.

Unfortunately, rural credit markets do not function as smoothly as implied here. Farmers may have difficulty in borrowing as and when they wish or in

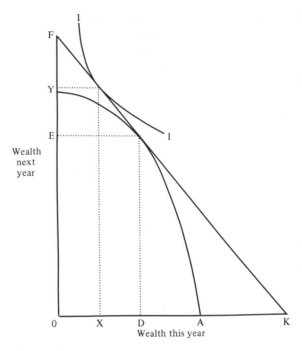

Fig. 8.6 Gains from lending

lending their savings out at interest. In any case risk has an important influence on the costs and volume of credit.

First, let us consider the borrower's situation. Generally, he must undertake to repay instalments and interest on the loan annually regardless of the state of the harvest. In good years he should produce a surplus over and above the cost of servicing the loan, but in bad years, even if he makes a loss, he still has the loan service charges to meet. His residual net income will therefore fluctuate more widely than if he had not incurred the debt. Indeed, the more that he borrows the greater will be the variation in his residual net income. This is known as the 'principle of increasing risk' and may be restated as follows. The greater is the ratio of borrowed to owned or 'equity' capital (the ratio being referred to as leverage or gearing), the larger is the risk. Clearly, for any risk-averse individual, this sets a limit on the amount he would wish to borrow.

However, the lender also faces a risk, that the borrower will default and the loan will not be repaid. Before making a loan he may require some reassurance that the borrower is creditworthy. This can be provided in two main ways:

(i) Personal knowledge of the borrower or even control and supervision of his operations; or

(ii) Collateral security, meaning some possession of the borrower which the lender can keep if the loan is not repaid. Under a mortgage, land or buildings provide the collateral security. The same is true when land and tree crops are pledged. Machinery or equipment bought under hire-purchase is itself collateral security in case the payments are not made. Many small farmers have nothing to offer as collateral security.

Interest rates charged on loans should cover

(i) the opportunity cost of the funds, that is the base rate of interest,

(ii) administrative costs, which are proportionately higher the smaller is the loan, the more remote is the borrower's location and the greater is the need for personal contact and supervision,

(iii) losses due to default; for which a 'risk premium' is included,

(iv) inflation, which effectively means a fall in the value of money.

These costs in combination may result in relatively high interest rates on unsubsidized rural credit.

Credit needs and sources

Despite the possible high rates of saving achieved by some farmers, there is little doubt that credit can improve the productivity, incomes and welfare of rural people. Short-term credit may alleviate seasonal needs for working capital, or the problems arising from crop failure, sickness within the family or unexpected social commitments. Note that credit may be used for production or consumption and some argue that it should be restricted to the former use. However, the distinction is really very hard to draw since funds are 'fungible' which means they can be moved around between uses. Credit provided for productive purposes may simply allow the family to spend more of their own savings or equity on consumption.

However, savings which are already invested in the farm or in non-agricultural activities are no longer fungible. A household with savings tied up in goats cannot immediately release these funds to buy sheep if the latter prove more profitable. Since practically all technological innovations are embodied in new forms of capital, liquid funds are needed to finance their introduction. The provision of credit will facilitate and accelerate the necessary investment and hence the adoption of the technology. However, in the absence of any new technology, provision of credit alone may have little impact on agricultural production.

Credit agencies are frequently classified into two groups; formal and informal. Formal institutions include banks and co-operative credit unions, while informal agencies may be further sub-divided into two groups. On the one hand, family members, kinsmen and farmers' credit associations (such as *esusu* clubs in Nigeria) lend money or assets at little or no interest. On the other hand, village traders and moneylenders sometimes charge very high interest rates. None the less, informal sources, even of the latter kind have certain advantages over formal institutions.

(i) They are convenient, available locally, require no documentation and can provide credit quickly.

(ii) The informal moneylender has local knowledge to help in appraising household credit needs and creditworthiness.

(iii) There is little risk of default because the lender is generally well placed to apply pressure on the borrower to ensure payment.

(iv) Administrative costs are low.

(v) With some types of informal credit such as the caretaking of livestock and the pledging of tree crops, the risks of borrowing are shared between the two parties involved.

The disadvantages of informal credit are that

(i) the borrower feels he has an obligation to the lender and loses his independence,

(ii) there are few alternative sources to choose from, and

(iii) only short-term and relatively small loans are available.

Formal credit institutions have standardized lending procedures and make loans on a contractual basis. Such agencies may make longer term and large loans available but they face special problems in lending to farmers.

  (i) Since farms are widely scattered and many are remote from urban centres and main roads, the costs of travel and administration are high. When rural branches are established they may do much less business than comparable town branches.

 (ii) Compared with urban industry, farmers' loans are relatively small. The cost of administration and supervision of a small loan is the same as for a large one.

(iii) Many farmers lack knowledge and experience of formal application procedures. They are discouraged by the need to complete application forms and other documents, which naturally introduce delays.

 (iv) Agricultural production is particularly risky. As a result risks of default are high.

  (v) Farmers lack suitable collateral security. Formal mortgaging of property is often not feasible because farmers do not have legal title to their land. In any case it may be difficult socially and politically, to foreclose on a smallholder's main productive assets (see Chapter 6).

As a result of these difficulties, formal credit agencies rarely operate in rural areas without government support and promotion. Farmers are often forced to rely on informal sources.

Promotion of rural credit

Four main topics deserve brief mention. These are (i) subsidized credit, (ii) lending through marketing agencies, (iii) co-operation and group loans and (iv) lending as an element of integrated rural development.

Subsidized credit is provided to farmers in many African countries, through their Agricultural Development Banks. At first sight this appears a straightforward method of assisting small farmers and promoting agricultural production at the same time. However, there are several possible disadvantages. First, the low interest rate may discourage farmers from saving themselves. Second, it cannot be self-supporting, a continued subsidy is needed, which means that the total amount of subsidized credit must

be limited. Third, since it cannot be rationed by price, that is by allowing the interest rate to vary, it must be rationed in some other way, usually on the basis of estimated creditworthiness. Fourth, as a result the bulk of loans go to the larger farmers, who since they have more equity to offer, inevitably appear more creditworthy. Finally, the large farmers who are able to get the loans, have less incentive to use the credit productively when it is cheap. Farmer surveys suggest that other factors such as ease of securing credit, timeliness of loans, transaction costs, and collateral requirements are more important in influencing use of formal credit, than the rate of interest charged.

Lending through marketing boards and similar marketing agencies is particularly suited to short-term loans offered at the start of the cropping season. Repayment is then deducted from the price of the crop after harvest. The system is somewhat inflexible and involves strict control of the marketing operation.

Group loans have some advantages over individual loans. Generally, since a larger sum is lent to the group, the per unit administrative costs are reduced. In addition, since the group takes joint responsibility for repayment, there is less risk of default. Thus considerable cost savings may be made. At the same time, co-operation in production and marketing may be encouraged through the joint action in raising the loan. Indeed, once a national structure of primary village co-operatives and credit unions has been established, this may serve as a banking system providing funds to farmers from the central co-operative bank. Farmers may also be encouraged to save and deposit their savings with the co-operative.

The main advantages of linking credit with Integrated Rural Development is that opportunities for profitable investment are presented along with the credit needed to implement them. Indeed, some of the credit may be provided in kind. At the same time agricultural extension agents may assist in drawing up plans and making credit applications. They may also be involved in supervision of its use and repayment. The provision of tractor and other hire services is another means of making more capital available to farmers. However, Integrated Rural Development Projects and tractor hiring services have often proved costly and of limited success in increasing agricultural productivity.

Further reading

Ahmed, I. & Kinsey B. H. (1984). *Farm Equipment Innovations in Eastern and Central Southern Africa*, Aldershot, UK, Gower

Hill, P. (1970). *Studies in Rural Capitalism in West Africa*, African Studies Series No.2, Cambridge University Press

Howell, J. (ed) (1980). *Borrowers and Lenders: Rural Financial Markets and Institutions in Developing Countries*, London, Overseas Development Institute

Hunt, D. (1975). *Credit for Agricultural Development*, East African Publishing House

Miller, L. F. (1977). *Agricultural credit and finance in Africa*, New York, The Rockefeller Foundation

Upton, M. (1962). Costs of maize storage 1959–60: the cost of guinea-corn storage in silos 1959–60. *West African Stored Products Research Unit: Annual Report 1962*, Ibadan, Nigeria

Upton, M. (1966). Tree crops: a long term investment. *Journal of Agricultural Economics*, **17** (1) p. 82

Upton, M. (1976). *Agricultural Production Economics and Resource Use*, Oxford University Press, Chapter 7

Upton, M. (1985). Returns from small ruminant production in South West Nigeria, *Agricultural Systems* **17**, p. 65

9

Management

Managerial efficiency

The role and functions of management were discussed in some detail in Chapter 1. It was argued that physical resources of land, labour and capital are not productive unless they are organized and co-ordinated by someone who makes the necessary decisions and carries them out. To this extent, 'management' may be viewed as a productive resource. Without inputs of management, a farm would not exist.

Within the agricultural industry, the main managerial task devolves upon the many smallholder families which make up the farming population. Governments also intervene to change the structure of the industry, to encourage the employment of hired managers or to improve the standard of management on family farms. Policy choices must be influenced, not only by considerations of returns to scale already discussed in Chapter 4, but also by assessments of the managerial efficiency of the smallholder population.

In an influential book, written over twenty years ago, (Schultz, 1964), it was argued that 'traditional farmers', are poor but efficient. This means that on farms, where *no* new technology or new resources have been introduced over a long period, so farmers have had time to adjust, all opportunities for improvement have been exhausted. Farmers then operate at, or near, the economic optimum where the marginal value product is equal to unit factor cost for all variable inputs (see Chapters 2 and 4). The policy implication is that increased output and farm incomes can only come about through technological innovation, since there is little scope for improvement by reallocation of existing resources. There is some

evidence to support this view, based on production function analysis of farm survey data (see Chapter 13). These studies suggest that the potential increase in farm income from equating the value of the marginal product of all inputs in all uses would generally be less than 10 per cent (see Welsch, 1965, Massell, 1967, Johnson, 1969 and Shapiro, 1977).

Many criticisms have been made of this theory and the methods of measurement used. To start with the empirical studies only claim to show that *on average* resources are allocated at, or near, the economic optimum point. But this fails to take account of the variation between farms; some must be operating below the economic optimum and some above. As a result there may be considerable scope to improve resource allocation on individual farms, even though this is not apparent on average.

A more fundamental weakness is the assumption that all farmers are operating on the same production function, or that they are all technically efficient. In practice, some farmers produce less output than others from a given combination of inputs; they are technically inefficient. The distinctions between allocative and technical efficiency are illustrated in Figures 9.1 and 9.2, which represent a production response curve, like Figure 2.3, and an isoquant, like Figure 2.6, respectively. However, it is now recognized that these curves represent the 'outer-bound production function'. The response curve of Figure 9.1 shows the *maximum possible* level of output for each level of input. The isoquant of Figure 9.2 shows the *least quantities* of the two inputs needed to produce the given level of output. Farmers operating at point A or point C on the curve are technically efficient, although the latter represents an inefficient allocation! However, other farmers, who operate at

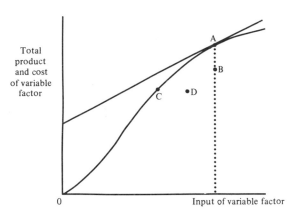

Fig. 9.1 The 'outer bound' production response curve

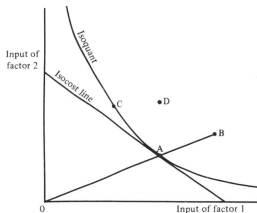

Fig. 9.2 The 'outer bound' isoquant

points B or D 'inside' the curve, are technically inefficient; they use more inputs to produce a given output, than their neighbours do.

A study of Tanzanian cotton farmers (Shapiro, 1977), involved identifying those farmers who attained the highest output : input ratios and were assumed to operate on the outer-bound production function. The practices these farmers were using, relating to fertilizer rates, numbers of sprayings and so on were closely in accordance with those recommended by the extension service. A comparison with the rest of the farmers, showed that only a small minority were operating at or near the 'technical optimum' represented by the outer-bound production function. Most of the farmers were *not* operating near this boundary, and it was suggested that by raising their technical efficiency to that of the 'best' farmers, total output would be 51 per cent higher.

Other studies have involved attempts to measure managerial ability; by applying the classification of master farmers, plot holders and co-operators used by the extension service in Zimbabwe (Massell, 1967, Johnson, 1969) or by calculating an index based on farmer attitudes and aptitudes in South West Nigeria (Upton, 1970) and Zambia (Tench, 1975). The results of these studies suggest that differences in 'managerial ability' do have some impact on productive performance. Further support is provided by the finding the farmers' yields are generally much lower than those obtained under experimental conditions on research stations. This may be due, at least in part, to poorer management on farms.

Thus we might conclude that, contrary to the poor but efficient hypothesis, there is some scope for improving farm productivity and incomes through better standards of crop and livestock husbandry. Furthermore, gains might be substantial if all farmers could achieve the standards of the most efficient. However, the scope may be limited for several reasons.

(i) Some variation in managerial skills is due to differences in natural ability, and cannot be learned.

(ii) Variation in performance may be due to unrecorded or unidentified differences in resource endowments, such as land quality, location relative to markets, sources of off-farm income and family age structure, rather than to differences in management.

(iii) Some differences in performance are due to good or bad luck, the incidence of pests and diseases on crops, livestock or family members, for instance.

(iv) Farmers differ in their attitudes and objectives.

Attitudes and objectives

This discussion of the allocative efficiency of traditional farmers has been strongly criticized on the grounds that they have other objectives than profit maximization. Economic efficiency should not be judged against this as the sole criterion. Another important objective, as we have seen in Chapter 5, is risk avoidance. A risk-averse farmer will forego some profit in exchange for greater security or improved chances of survival. The 'optimum' for a risk averse farmer occurs when the marginal value product of a variable input is *greater than* its unit cost. It is sub-optimal in terms of the profit objective.

Furthermore, individuals or families differ in their attitudes to risk. The more cautious are willing to forego more profit in exchange for greater security, than are their more adventurous neighbours. Clearly, these attitudes are likely to depend upon background and environment. Where the physical environment makes agriculture difficult, as in the semi-arid zones, and where the fear of natural catastrophe is always present, farmers are likely to be more cautious than those in high potential areas. Similarly, the family with few resources relative to household needs is likely to be more risk averse than wealthier neigh-bours.

Another defect of the concept of poor but efficient traditional farmers, is that very few traditional farmers exist in the strict sense used above. Wide-spread changes are occurring in population densities and man–land ratios, in market opportunities and in available technology. Agricultural development requires a willingness by farm families to adapt and innovate. But this is related to risk attitudes. An innovator is one who is willing to take risks in early adoption of new products or practices. A late adopter or laggard is more cautious. In the past, it was common practice for the extension service to seek out the farmer innovators and assist them in leading the diffusion of innovations. However, this approach may foster differentiation and inequality between innovators and others, so a more general group approach to agricultural extension is increasingly favoured.

Of course, farmers have other objectives than risk avoidance or survival, which really makes the mea-surement of whether they are 'efficient' in attaining their objectives very difficult indeed. It may be argued that small farmers are innovative and rela-tively successful in attaining their objectives, given their limited resources (Richards, 1985). This raises the question of whether there is scope for improving

farmers' managerial skills through training or exten-sion programmes aimed specifically at that end. Gen-erally speaking, it is doubtful whether such pro-grammes can be justified except where major and significant changes are being introduced, say from pastoralism to irrigated agriculture or from hoe culti-vation to tractorization. Agricultural productivity is more likely to be increased by institutional change which improves farmer access to resources, improved market incentives and above all technical research and development.

The farmer's background and environment

We should perhaps remind ourselves at this point of the complexity of decision-making in a farm house-hold. Responsibility for some decisions rests with the head of the household, who may be male or female, some decisions are shared, while others devolve upon other family members who manage their own plots, livestock or off-farm activities. When we talk of the farmer's personal characteristics we are really con-cerned with whoever is responsible for decision-making. For situations where decisions are made jointly, our discussion must relate to the character-istics of the decision-making group.

Managerial ability is only one of the farmer's per-sonal characteristics which directly influences the decisions made. Others are summed up under the headings of attitudes and objectives and background and environment. The 'management process' is sum-marized in Figure 9.3. Arrows representing the decision-making process lead from left to right but the feedback loops leading from right to left should be noted.

The background and environment includes the resources at the farmer's disposal as well as the social and cultural constraints influencing his choice. Environmental factors influencing farming systems and their performance were discussed in Chapter 1. Many of these features are similar for all farms in a given region. The climate, soils and vegetation are likely to be fairly uniform within a particular agro-ecological zone, land resources are generally allo-cated equitably, markets for inputs and outputs are the same for all producers while there may be certain socially accepted norms of behaviour. However,

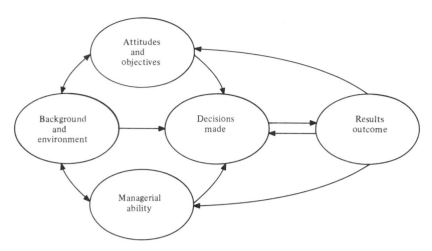

Fig. 9.3 The management process

there are also differences between individual households in the bundle of resources they command and in the market opportunities they face. Local differences in soil type and location relative to markets, must influence the decisions that are made and their outcome in terms of productive performance.

Education has an important influence on managerial ability and decision-making. Used in its broadest sense, education must include not only formal schooling but also agricultural extension and, indeed, any experience which broadens a person's outlook and increases knowledge. Education may contribute to

(i) deeper understanding of the environment and current productive activities;

(ii) greater awareness of new opportunities for improvement, including new technologies and new market opportunities;

(iii) more technical skill in managing both existing and new productive activities;

(iv) improved managerial ability in establishing objectives, decision-making, taking action and controlling operations.

Thus differences in educational background may account for differences in production performance between farmers.

Age and stage in the family life-cycle also influences managerial decision-making. On the one hand, as a farmer ages he gains experience, which should result in improved managerial ability. On the other

hand, the resource base changes. Early in the cycle when children are small, the dependency ratio is high and labour resources scarce. As children grow up and contribute to the work force, the farm can grow. In old age the ratio of dependants to workers may rise again and productive activity must be curtailed. The labour situation may be particularly difficult in female-headed households (see Chapter 7).

However, managerial ability together with good or bad luck also influences the amounts of resources controlled as the outcome of past decisions. The more technically efficient manager is likely to control more resources than his less efficient neighbour. This may be illustrated theoretically using Figure 9.4, which shows the production response and marginal product curves to hired labour for two farmers A and B. Farmer A is supposed to be more technically efficient than farmer B, as a result either of better management or better fortune. Thus for any level of labour use the marginal product is lower for farmer B than for farmer A. If we assume they both pay the same wage rate and receive the same price for their product, and that they are both profit maximizers, the economic optimum for farmer B is below that for farmer A. This, in turn, means that farmer A can clear more land and accumulate more savings than farmer B. Thus the more successful manager can attract resources away from the less able and differences in wealth may grow, with commercialization of agriculture.

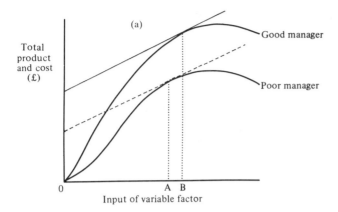

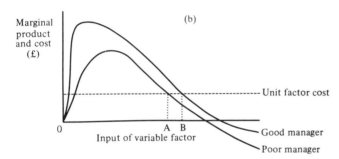

Fig. 9.4 Management and scale, (a) Total product,
(b) Marginal product

This may be exacerbated by the increased use of credit, since formal agencies are likely to find the 'better' manager more creditworthy than his less technically efficient neighbour. The influence of good or bad management is therefore not just a constant difference in performance, but a dynamic, cumulative and growing discrepancy in resources controlled and income earned.

Project management

The term 'project management' is used here to identify the management of large-scale development projects, whether they are essentially commercial production schemes or some form of integrated rural development. In either case the role of management is very different from that on family smallholdings. This is largely due to the hierarchical administrative structure within which a project is operated. Thus the project manager is responsible *for* a large number of subordinate staff and is responsible *to* his superiors either in government or the owners of commercial enterprises in the private sector. Some of the tasks involved are illustrated in Figure 9.5 (after Wiggins, 1985).

Project management is not the main subject matter of this book, although the principles discussed here have considerable relevance (see Smith, 1984). However, we will consider two related issues, one being the training of project managers and the other their allocation. It should be clear from a brief consideration of the tasks of project managers (Figure 9.5) that training is needed. The usual qualification required is a University degree in agriculture or some allied subject. But this is not necessarily the most appropriate form of training for the task. Short courses at agricultural or management training colleges which involve more practical experience and case study exercises may provide better preparation.

The other consideration is that, since project managers require special skills and training, they form a separate and distinct resource from smallholder farmers. The former are scarce and costly. It is therefore highly desirable that trained managers are employed as effectively as possible.

There is a clear association with economies and diseconomies of scale, (discussed in Chapter 4). One of the main diseconomies of large-scale agriculture is the high cost of the skilled management which is needed. On larger schemes the managerial hierarchy grows both vertically (the number of different levels of management) and horizontally (the number of managers at each level). Thus the costs escalate at an increasing rate with increased size. This in itself may outweigh the technical benefits of large-scale production in some cases.

There are two alternative approaches to the organization of large-scale projects. One is to impose strict control and rigid rules of behaviour for everyone involved. This is broadly the approach used on many plantations, state farms and some irrigation schemes. The alternative is to allow considerable autonomy and freedom of action to individual households, as may be the case on co-operative settlement schemes. In this case, careful planning and organization may be needed to provide appropriate incentives to

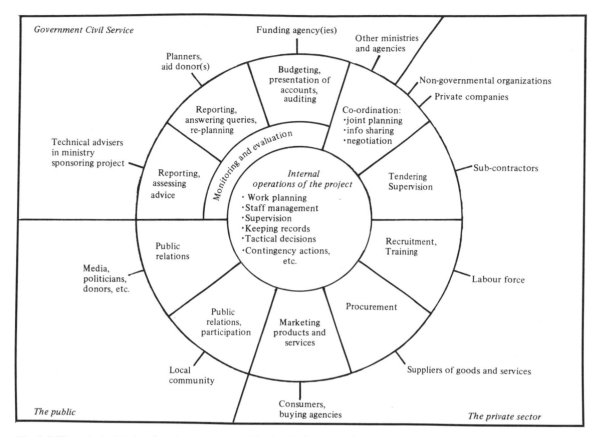

Fig. 9.5 The principal tasks of project management in developing countries

participants. However, if it can be organized, this approach can result in substantial savings of trained managerial manpower. Fuller use is made of the managerial skills of the rural households. Even more effective use of trained managers might result from abandonment of the large-scale transformation approach, provision of more autonomy to small-holder families and employing trained managerial staff as advisers or Farming Systems Researchers.

Further reading

Johnson, R. W. M. (1969). The African village economy: an analytical model, *The Farm Economist*, **11**(9), 359

Massell, B. F. (1967). Farm management in peasant agriculture: an empirical study, *Food Research Institute Studies*, **7**(2), 205

Moock, P. R. (1981). Education and technical efficiency in small-farm production, *Economic Development and Cultural Change*, **29**(4), 723

Richards, P. (1985). *Indigenous Agricultural Revolution: Ecology and Food Production in West Africa*, London, Hutchinson

Schultz, T. W. (1964). *Transforming Traditional Agriculture*, New Haven, Yale University Press

Shapiro, K. H. (1977). Efficiency differentials in peasant agriculture and their implications for development policies, in *Contributed Papers Read at the 16th International Conference of Agricultural Economists*, Institute of Agricultural Economics, Oxford

Shapiro, K. H. & Muller, J. (1977). Sources of technical efficiency: the roles of modernization and information, *Economic Development and Cultural Changes* **25**(2), 293

Smith, P. (1984). *Agricultural Project Management*, Elsevier Applied Science

Tench, A. B. (1975). *Socio–economic Factors Influencing Agricultural Output: With Special Reference to Zambia*, Saarbrucken, Verlag der SSIP-Schriften

Upton, M. (1970). The influence of management on a sample of Nigerian farms, *Farm Economist,* **11**(12), 526

Welsch, D. E. (1965). Response to economic incentives by Abakaliki rice farmers in Eastern Nigeria, *Journal of Farm Economics,* **47**, p. 900

Wiggins, S. L. (1985). *The management of rural development projects in developing countries*, Development Study No. 27, University of Reading, Department of Agricultural Economics

PART III

Field investigations

10

Farming systems research

The approach

Farming systems research (FSR) is aimed at identifying options for improving the wellbeing of rural households in specific local environments. Much of this research has been conducted by staff of the International Agricultural Research Institutes, mentioned in Chapter 1, with the prime objective of developing new, improved farm-level technology. However, the introduction of new technology is not the only way of improving the wellbeing of rural households. Other possibilities include the provision of rural social infrastructure such as roads, marketplaces, health and education facilities, water and electricity supplies and opportunities for off-farm employment. Assured supplies of farm inputs and markets for farm produce and improved price incentives also benefit farm families. In principle, farming systems research may be used to identify options for improvement in all these areas.

There are four main characteristics of farming systems research. First and foremost it is focussed on the farm household. Thus it is based on a recognition that rural change and development ultimately depend on rural people whose existing practices are well adapted to environmental constraints and household objectives. Attempts to develop and impose innovations or policies from the top down, without previous reference to those who will be affected, rarely succeed. Those of us, research scientists or development planners, who are concerned to promote rural development must work with farmers and try to understand their aims, their methods and their problems if our outside assistance is to be acceptable and useful to them. Furthermore, proposed innovations should be subjected to on-farm testing to identify practical management problems or constraints on their adoption before introducing them more widely.

The second characteristic follows from the first: namely that FSR is locale specific. Since its central concern is with farm households and since there are large differences between localities in the resource base and the farming systems practised, each FSR programme relates to a limited number of similar farms in a given locality.

The third main characteristic is that FSR is holistic, which means that it is concerned with the whole system and its interdependencies rather than with individual elements. In this respect it may be contrasted with commodity programmes which are aimed at increasing output of a single crop or livestock product. This single commodity approach fails to consider the repercussions a new product or process will have on the rest of the farm household system. To the commodity specialist, 'every isolated improvement is another brick in the building of a more efficient agriculture. In reality, however, every improved technique affects the whole structure of the farm. Its introduction does not represent the laying of a brick on top of a building, but the removal of one part way down and replacing it by a better one. This replacement can be as disturbing to a farm as to a building'. (Jolly, 1957). Although commodity programmes may be more effective in generating new technology they should always be complemented by FSR.

However, this argument begs the question of where the boundaries of the whole system lie. Much FSR has concentrated exclusively on the farm or even on a subsystem as in the case of Cropping Systems Research. Given that the central concern is with the wellbeing of the rural household, off-farm activities should be included in the system being studied. Links with rural services such as credit, input delivery, product markets and agricultural extension are also important and should form part of the study. Pastoral Systems Research, which is also included under the broad heading of this Chapter, incorporates (i) rangeland ecology, (ii) livestock husbandry, and (iii) the study of pastoral society and the household economy. Here again external trade is likely to have an impact on family welfare and should be included in any study of pastoral systems (see Upton, 1986).

The fourth main characteristic is that FSR is multidisciplinary. In particular it integrates the perceptions of both the technical and the social sciences to analyse existing systems and to identify options for improvement. Technical sciences are needed to explain the physical relationships between inputs and outputs and to develop new products and new methods. The social sciences can contribute to understanding how the society is organized, how resource allocation decisions are made, how the disparate parts of the system are integrated and which new products and new methods are likely to be accepted. Within these two broad areas there are many scientific disciplines that could contribute as suggested in the opening chapter. One essential discipline, however, is agricultural production economics or farm management.

Thus FSR usually involves multidisciplinary teams of researchers. The team size, and hence the number of separate disciplines represented, is limited by cost considerations, frequently to one crop scientist and one production economist or farm management specialist. No matter how large or small the team, FSR requires a flexible open-minded outlook by team members. Each one must be able and willing to think in disciplines outside his own and to learn from his colleagues as well as from farmers and their families. The, perhaps unattainable, ideal farming systems researcher would be competent in more than one discipline encompassing both technical and socio–economic knowledge.

Research procedures: description and diagnosis

There are four main stages involved in a FSR programme. These are;

(i) description and diagnosis,
(ii) design of improved systems,
(iii) testing and evaluation of improved systems, and
(iv) implementation and extension of promising alternatives.

Although most farming systems researchers agree with the general philosophy outlined above and that these are the four main stages, their views differ considerably with regard to detailed procedure and indeed the general scale and timing of these stages. While Collinson, working in East Africa, suggests that an FSR study can be completed within three months, 'including a two-week input from the relevant technical scientists', others such as Norman who worked in Northern Nigeria operated on a time span of several years (see Collinson, 1979, Norman, 1980). In fact, the general philosophy and the four main stages may be executed at different levels. On the one hand, the entire research programme of an institute may be planned for several years ahead along these lines. On the other hand the procedures may be used for a series of relatively short studies in different localities within a broad agro-ecological zone. Figure 10.1 illustrates how such a series of studies might be linked together and with the experimental station activity. The four main stages will now be discussed in more detail.

The first step in description and diagnosis is to identify the target population of farmers the research is aimed at. This involves grouping farmers with similar, though obviously not identical, farming systems into zones or 'recommendation domains'. The target population for a particular FSR study consists of all farm families within the selected recommendation domain. A considerable amount of local information is needed and some preliminary investigational work may be involved before recommendation domains can be identified. There is therefore an overlap with the next step which is the assembly of background information. This includes agro-ecological data such as monthly rainfall, topography and soil types which determine the crops that can be grown and may be critical constraints on the system. It also includes measures of population density and, where available, production and yields of major crops and livestock. Information on local social structures and institutions, markets and prices for inputs and out-

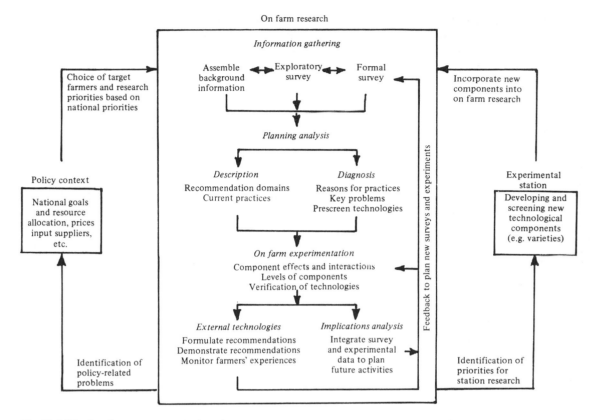

On farm research

Information gathering

Assemble background information ⟷ Exploratory survey ⟷ Formal survey

Choice of target farmers and research priorities based on national priorities

Incorporate new components into on farm research

Planning analysis

Description
Recommendation domains
Current practices

Diagnosis
Reasons for practices
Key problems
Prescreen technologies

Policy context

National goals and resource allocation, prices input suppliers, etc.

Experimental station

Developing and screening new technological components (e.g. varieties)

On farm experimentation
Component effects and interactions
Levels of components
Verification of technologies

Feedback to plan new surveys and experiments

External technologies
Formulate recommendations
Demonstrate recommendations
Monitor farmers' experiences

Implications analysis
Integrate survey and experimental data to plan future activities

Identification of policy-related problems

Identification of priorities for station research

Fig. 10.1 The farming systems research cycle

puts and government policies is also relevant as all these factors may influence or constrain what is produced. Most of this background information can be obtained from secondary sources, such as published reports, unpublished records and experienced local government officers or development agents.

More detailed description of farming systems and diagnosis of limiting constraints requires field investigations of the target population. There are wide differences of opinion as to the appropriate investigational methods to use. Some authorities recommend 'rapid rural appraisal' based on a tour of the zone, purposely visiting areas remote from the main roads, and in-depth interviews of a few representative farmers. Others argue that such details as the seasonal pattern of labour inputs, needed for the identification of seasonal labour constraints, can only be obtained by regular farm visiting over at least one cropping season. Opinions also differ as to whether

reliable data can be collected from a few case studies or whether a random sample survey of a large number of farms is needed. Clearly, the so-called 'quick and dirty methods' save time and costs in comparison with full-scale socio–economic surveys and may allow direct contact between researcher and farmer rather than through enumerators. Yet there are dangers of error and bias in relying on non-random samples and on farmer recall. These issues are discussed in more detail in the next Chapter.

Having collected information from farm families, the next step is its analysis in order to identify their objectives and preferences and the diagnosis of their problems arising from resource constraints, and environmental risks. This diagnosis should indicate points in the system where the introduction of new technology or the modification of government policies will have the greatest impact. It should identify where change is needed or where there is a potential demand for change. The methods used will generally include accounting procedures to summarize results for individual households and statistical procedures for aggregating and comparing these results to draw conclusions about the whole population. Some methods of analysis will be discussed in Chapters 12 and 13. Planning methods such as linear programming (Chapter 16) may also be used to model the existing system and diagnose constraints.

Research procedures: design, testing and extension

The scale and duration of the design stage depends upon whether appropriate technologies have already been developed and are available, on the shelf, to be taken down and used when needed. If this is not the case, then substantial scientific and technical research programmes may be required to produce new technology. In fact, there is no guarantee that the desired technology will be developed within a given budgetary and time allocation. The outcome of any research programme cannot be predicted with any degree of certainty so the supply of new technology will not necessarily match up with the demand. There are no grounds for assuming that FSR is a panacea which will *ensure* the generation of appropriate new technology. However, it should help to establish whether the technology that is generated is appropriate.

To this end the development of new agricultural technology may be set in an FSR framework even though it may be based on an experimental station. Such programmes are described as 'upstream FSR', to distinguish them from farm-based or 'downstream' programmes, although a more apt description might be 'resource management research'. The research on minimum tillage systems at IITA has hitherto been of this type.

The design of new improved systems may be assisted by the use of planning tools such as partial budgeting, or investment appraisal. In some circumstances more sophisticated models such as linear programming may be appropriate. Although time, effort and skill are involved in building models, they are potentially very useful in allowing the assessment of the impact of new technology and other changes before they are implemented. In short, models may be used in place of real farms to explore the effects of proposed changes on an experimental basis. It is generally much cheaper to experiment with models than with real-world farming systems. The main planning tools are described in Part IV of this book.

On-farm testing and evaluation is an important stage in FSR. This involves the introduction of the new improved system on a sample of farms with a view to measuring its feasibility, performance and impact in practice. Farmers may introduce modifications and improvements to the system during the testing phase. Careful monitoring is needed over at least one season in order to make an evaluation. Here again rigorous measurement of the advantages of the new system requires monitoring a representative sample of adopters as well as an otherwise similar sample of non-adopters. Statistical methods may then be used to test whether differences between adopters and non-adopters are significantly greater than zero. However, in practice, time and resource constraints may limit testing to a small number of case study farms.

Once a new system has been tested and shown to be an improvement on existing methods, the production stage may be launched. This involves promoting the spread of the new system throughout the target population. As local agricultural extension agents are generally responsible for this stage, it is highly desirable that they should be involved throughout the FSR programme. There should be a constant flow of information and knowledge between farmers, researchers and extension staff. The feedback of information regarding farmers' attitudes to the innovation and the managerial adjustments they make may be useful in designing future programmes. Reliable delivery systems for the necessary inputs and markets for new products are essential preconditions for the success of the production stage and the widespread adoption of innovations.

Issues of farmer involvement

There are several unresolved issues relating to FSR not least of which is the question of farmer involvement. Despite widespread agreement that new technology should be appropriate to rural household needs and compatible with existing farming systems, there are different schools of thought regarding the extent to which farmers should be involved in the research and development process. One school favours commodity programmes and upstream FSR. They claim that this is the most promising approach to making radical new discoveries and achieving large and significant improvements. In their view the downstream farming systems researcher, by concentrating on what farmers are currently doing, takes too narrow a view of what is possible and will probably only identify minor or marginal improvements, (Eicher & Baker, 1982 op cit.).

Following this approach, social scientists based at the research station are engaged in evaluating ongoing research and conducting field investigations. These are often based on full-scale random sample survey of farm families, with formal questionnaires, regular visiting and possibly even direct measurement of plot areas, yields and dietary intake for instance. If the aims can be achieved, such a survey should provide accurate and reliable estimates of the main variables measured. However, it is costly, time consuming and necessitates the use of enumerators. Thus direct contact between researcher and farmer is limited. Although the survey is focussed on the farm household system, farmers are not active participants in the research and development process. Similarly, where the agricultural extension service takes sole responsibility for communication with farmers there is little scope for direct interaction between researcher and farmer.

The informal, 'quick and dirty' survey, on the other hand, may be carried out by the researcher himself with the aid of an interpreter where necessary. Thus there is direct contact between research scientist and farmer. Although a checklist of important issues may be used, there is no formal questionnaire so interesting and relevant issues can be pursued in great depth and detail. The farmer is then a more active participant in the study. Indeed, he may frequently lead the discussion. Supporters of this approach suggest that a much deeper understanding of the farmers' attitudes, objectives, constraints and indigenous technology may be acquired than is possible using more formal methods. However, it may require a change in attitude by the research scientist towards a greater willingness to accept and learn from farmers' opinions. Special skills, derived from field experience or training, are needed to arrive at a full understanding of a farming system on the basis of a single interview. Some critics of this approach doubt whether it is possible to identify farmers' objectives simply by asking questions. On-farm testing of new technology is another element of this approach to farmer involvement.

Another school of thought presses the case for farmer participation even further. According to this view, a great deal of 'informal sector R & D' (research and development) already occurs on farms and in villages. Many successful innovations have been developed in this way without outside intervention. Thus there is a case for 'participatory research' much of it undertaken by farmers themselves. 'The role of the scientist is that of consultant: to collaborate rather than direct.' (Richards, 1985 op. cit.). This, it is argued, is an efficient way of meeting localized research needs and of mobilizing local skills and initiative. Problems that might be associated with the widespread adoption of this approach are first, that of organizing and administrating effective farmer panels, secondly the dangers of creating an élite class of 'panel farmers' and third, the training of scientists to adapt to this kind of work.

Clearly, there is a wide range of opinion regarding the appropriate level of farmer involvement. However, the different approaches are not all mutually exclusive. Thus Collinson recommends that informal diagnostic pre-survey by research scientists should be followed by a formal survey of farmer circumstances (Collinson, 1979). On the other hand, scientists working in teams on a research station cannot easily operate as consultants to local farmer panels. Unfortunately, it is very difficult to evaluate and compare these approaches objectively. Apart from the problems of identifying and measuring all the costs and benefits of any FSR programme, comparisons may not be valid because the success of a particular approach is likely to be influenced by the type of farming system, the natural and the socio–economic environments, the type of innovation developed, and the personalities of the individuals involved. All these key factors differ from one case to another. Thus, the choice must depend on personal judgement of what is appropriate to the particular circumstances.

Other unresolved issues

One of the main justifications for FSR is that it provides a focus or aim for agricultural research, the aim being to overcome constraints and improve the wellbeing of a specific group of farmers. Since research is directed to a specific purpose it should become more cost-effective; that is more effective for a given total cost or cheaper for a given level of achievement. However, FSR is itself a costly and time-consuming activity. Major unresolved issues relate to the possibility of cutting the costs of FSR without reducing its effectiveness. Time savings are particularly important since farming systems and their environments are changing. The research is aimed at a moving target. The results of FSR programmes, which take several years to complete, may be out of date by the time they are produced; they may miss the target. Thus if 'quick and dirty methods' which cost less time and money are as effective in identifying productive innovations as more rigorous and detailed research, they are clearly to be preferred. In practice, there may be a trade-off between cost and effect.

Another issue where this trade-off is important concerns the choice of recommendation domain. Costs may be reduced by restricting the study to a small group of closely similar farms. However, the effects or the benefits may also be restricted to this small group. Ideally the results of research should be widely adaptable over a very large area and a large number of farmers. While this may be possible with a commodity programme, development of a new high-yielding variety for instance, FSR is, by its very nature, locale specific. Thus difficult decisions must arise in striving for as large a recommendation domain as possible whilst ensuring that the population is reasonably homogeneous.

Finally, we return to the question of how holistic or comprehensive FSR should be. Costs can be saved by limiting the study to particular sub-systems; most have concentrated on arable cropping systems while permanent crops, supplementary livestock and off-farm work have largely been ignored. Costs are also saved by limiting the number of scientific disciplines involved. But these savings are made at the cost of abandoning the basic principles of a holistic and multidisciplinary approach. Ideally, farming systems should be related to the wider social environment, which means that village level and national policy studies should be incorporated into the FSR programme, but once again cost considerations may preclude this.

A closely related issue is whether FSR can, or should, be used in planning and implementing government policies for agricultural pricing, institutional change and rural development. We have argued that an understanding of farming systems, objectives, constraints, costs and returns is needed for all effective policy-making. It is not clear whether the FSR approaches outlined above are appropriate for providing all this information or whether separate surveys and studies are needed.

Further reading

Byerlee, D., Collinson, M. P., Perrin, R., Winkelmann, D., Biggs, S., Moscardi, E., Martinez, J. C., Harrington, L. & Benjamin, A. (1980). *Planning Technologies Appropriate to Farmers: Concepts and Procedures,* El Batan, Mexico, CIMMYT

Collinson, M. P. (1979). Micro-level accomplishments and challenges for the less developed world, in Johnson, G. L. & Maunder, A. (eds) *Rural Change: the Challenge for Agricultural Economists.* Proceedings, Seventeenth International Conference of Agricultural Economists, Banff, Canada; England, Westmead

Gilbert, E. H., Norman, D. W. & Winch, F. (1980). *Farming Systems Research: A Critical Appraisal.* MSU Rural Development Paper 6, Michigan State University, East Lansing, Michigan

Jolly, A. L. (1957). The unit farm as a tool in farm management research, *Journal of Farm Economics,* **39**(3) p. 739

Norman, D. W. (1980). *The Farming Systems Approach: Relevancy for the Small Farmer,* MSU Rural Development Paper 5, Michigan State University, East Lansing, Michigan

Upton, M. (1986). Production policies for pastoralists: the Borana case, *Agricultural Systems,* **20**(1) p. 17

11

Surveys

Purposes and methods

Farming Systems Research involves two kinds of farm household investigation; the descriptive and diagnostic study and on-farm testing. The first is aimed at description of the farm and household system, estimation of underlying relationships, specification of household objectives and diagnosis of key constraints and weaknesses. The second involves comparison of a group of farms which have adopted the innovation with a control group which have not, in terms of the impact on household wellbeing. In either case, many items of information are needed; it is a multi-subject enquiry.

Other studies needed for planning purposes may appear to have a simpler purpose such as estimating the average cost of production per tonne of maize on a particular type of farm. Even in cases such as this the simplicity may be more apparent than real. Not only are there many items of cost to be considered, but also, if the full opportunity cost of the resources used is to be estimated, some analysis of the whole farm household system is needed.

Thus all farm household studies are multi-subject enquiries. Furthermore, it is difficult to be precise about what we need to know or which items of information are essential. However, the following categories of data are generally needed.

1. Descriptive material on farming systems

This includes not only the areas and combinations of crops grown together with the seasonal cropping pattern and sequences, but also numbers of each class of livestock and the methods of production used. It may also be appropriate to describe associated off-farm activities.

2. Resource endowments

It is generally useful to know the resource base or the quantities of resources controlled by a typical household. This requires estimates of (i) the area of land controlled, (ii) the total family labour force and (iii) physical productive assets owned, such as livestock, permanent crops, tools and buildings. In addition, information is needed on the scope for acquiring more of these resources, extending the farm area, hiring labour or obtaining credit.

3. Input–output data

These are measures of the quantities of resources actually used as inputs and the physical yields obtained. They must be related to a given production period usually taken to be a year. However, more detailed information may be required on the seasonal spread within the year of labour use for instance. To obtain such information accurately may require fairly continuous observation or recording. Problems of measuring inputs and outputs under mixed cropping are discussed later.

Some productive activities, particularly permanent crops continue over many years. The pattern of annual inputs and outputs is likely to vary over the lifetime of such investments (see Chapter 8). Ideally, the whole input–output profile over the life of the investment would be measured, but the fact that most field investigations cover only one year precludes this. It may, however, be possible to record inputs and outputs for a given permanent crop on different plots established at different times. For example, inputs and outputs on cocoa nurseries may be separated from inputs and outputs for mature trees.

In such a way a time profile of inputs and outputs might be built up.

4. *Purchases and sales*

These data are needed for two purposes: one is to evaluate the financial position and cash income of the farm household, the other is to provide price data for evaluating all inputs and outputs. For some purposes it may be sufficient to know aggregate costs and returns, while for others a breakdown of these totals, by enterprise, may be needed. Clearly wages and remittances from off-farm occupations are included.

5. *Farmers' attitudes and objectives*

As already emphasized an understanding of farming systems and farmers' behaviour requires information on farmers' attitudes and objectives. This information would include attitudes to risk, to food self-sufficiency, tastes and food preferences, leisure and off-farm work requirements as well as social customs and taboos.

Warnings are often given regarding the dangers of collecting unnecessary data. Where, as in farm household investigations, there are multiple objectives, these should be ranked in order of priority. Unnecessary or trivial data should be omitted from the study to limit costs and facilitate data collection and analysis. Problems arise, however, when we are unsure as to whether a particular piece of information might be useful. It may be very costly to go back, after the main study is completed, to recover some critical datum item which was omitted from the main study.

Some information from each of these five main categories listed above is likely to be needed but it is impossible to generalize regarding the precise requirements. The importance of data on purchases and sales depends upon the farmers' attitudes to cash and subsistence farming. The need for detailed records depends upon whether labour is thought to be a critical constraint on production. Thus some prior knowledge of the system is needed in order to determine what data to collect. This lends support to the idea, adopted in much Farming Systems Research, of carrying out an area familiarization study using rapid rural appraisal before embarking on a more detailed formal survey. The preliminary investigation permits more precise specification of just which data items are needed from the formal study.

Data collection methods

There are three main methods for collecting farm household data, which are in order of increasing cost
(1) records kept by respondents;
(2) interviewing respondents;
(3) direct observation.

Records kept by respondents Where farmers keep formal records and accounts, these provide an ideal source of farm management data. In such circumstances a postal survey may be possible, thus eliminating the costs of enumeration. However, farm accounts are only likely to be kept on large-scale commercial or state farms and estates. This approach has little relevance to the vast majority of small farms. Early studies in Kenya relied on literate children to keep farm records and accounts for survey purposes (MacArthur, 1968) while 'emergent farmers' in Zambia were able to provide bank statements of their financial position (Bessell *et al.*, 1968) but such cases are atypical of the majority of farmers.

Interviewing respondents This is the usual method of investigating attitudes and objectives, and may be used for collecting factual information on farming systems, resource use, crop and livestock yields and research constraints. It is likely to require less frequent visiting and to be less costly than direct observation and measurement, but may produce some inaccuracies or biases.

Attitudes and objectives are described as 'latent variables' existing in the individual's mind but not necessarily easily expressed. Very few of us could specify precisely what is our aim in life in response to a simple question. There is a temptation to give answers which will satisfy or please the interviewer, rather than carefully exploring one's own motives. With regard to factual information, there is the problem of recall. Clearly, this is a possible source of inaccuracy or error. We return later to the question of frequency of visiting and errors of recall. However, there are possible advantages in relying on farmer recall when there is substantial year-to-year variation in the weather, resource use and yields. The study period may well be atypical in some sense, so that data collected by direct observation will also be atypical. The farmer's estimates of resource use and yields may be inflenced by his judgement of what is average rather than by what has occurred in the current season.

We may remind ourselves at this point of the possible difficulties in defining the basic unit of analysis: the household and the farm. There are difficulties in deciding exactly who should be included in the household in terms of both their contribution to household resources and their dependence upon household income. There are difficulties in identifying who makes the decisions and therefore who should be interviewed regarding his attitudes and objectives. In some cases decisions are made jointly by household members and group interviewing is more appropriate than individual questioning. There may also be difficulties in recording all the resources under the family control. Distant plots of land, areas under bush fallow and herds of livestock grazing far afield may easily be overlooked. Some authors have argued that the household is too small a unit to capture the multi-dimensional relationships affecting decision making on African farms. (Ancey, 1975 or Gastellu, 1980). Arguably the whole village or lineage should be the basic unit of investigation.

Direct observation This clearly involves regular visiting by the investigator or his enumerator and is therefore very time-consuming and costly. However, if it is done properly the results should be accurate and reliable. Clearly, it is impracticable to follow every member of the farm household all the time and record their every movement, besides observing crop and livestock growth and development. Hence direct measurement is always used in conjunction with interviews, to collect missing data. Direct estimations can be made of land areas and the resource stock with periodical measurement of labour use, crop yields and other input and output flows.

The three main types of field investigation are (1) case studies; (2) farm surveys of the rapid rural appraisal kind and (3) the cost-route method (Spencer, 1972). These are distinguishable in terms of (a) the number of farms involved and (b) the frequency of visiting. All three methods have been used in Africa.

Farm case studies relate to a few farms which are studied in great depth with regular visits, observations and possibly record keeping. Clearly whole village studies must be limited to very few cases, but some farm household studies have been of this nature (eg see Clayton, 1961). Unit farms, which are case studies established by the researcher, often on a research station, have been used to provide data and for on-farm testing in various parts of Africa; at the International Institute for Tropical Agriculture, Ibadan, for example.

Rapid rural appraisal is based mainly on interviews and informal observation. It involves few visits to each household, possibly only one, so the cost per household is relatively small and a larger sample can be covered for a given total expenditure than using the cost-route method. This approach is increasingly favoured because of its low cost and the advantages of completing a study within a short period of a few months (see Collinson, 1982; Byerlee *et al*, 1980). By contrast case studies or the cost-route method usually involve record collecting over a period of at least twelve months often with a similar additional period devoted to analysis and presentation of results. The greater timeliness achieved with rapid rural appraisal is a major advantage in providing data which are still relevant in a rapidly changing situation.

The cost-route method refers to repeated visiting of the same sample of farms over an extended period to collect data on inputs and outputs, costs and returns, some by questioning and some by direct observation. It is generally claimed that this method provides the most accurate and reliable data particularly for items such as labour use and crop yields. However, the cost per household of regular visiting is substantial. There is therefore an important trade-off between sample size and visiting frequency for a given total expenditure.

Summary statistics

The information collected from a farm household survey may be quantitative: areas of land, hours worked, or kilogrammes of grain for instance, or qualitative as in response to questions regarding attitudes. It may be further categorized in terms of the number of possible response classes. Thus we may identify

(1) binary data with only two response classes, such as whether the household head is male or whether any permanent crops are grown;
(2) multiple category data where there are a number of discrete categories:
  (a) non-numerical and unranked data, as in a set of alternative farmer objectives;
  (b) numerical or ordered data; such as the number of ox teams owned, the number of the month of planting or soil quality;
(3) continuous data on plot areas, crop yields or length of time worked.

For most practical applications we need to summarize the data, and different summary measures are suggested for each of the above categories. In the first case the appropriate summary measure is the *proportion* of positive responses. For category (2), the *mode* or most frequently occurring response may be used. Indeed if continuous data are grouped into classes, a *modal class* may be identified as the most frequently occurring class. However, for numerical or ranked data, whether discrete or continuous, the most common measure of central location is the *arithmetic mean*, or simply the mean. This is defined in the same way as the 'expected value' given that each observation is assumed to have a probability of 1/n where n is the total number of observations or sample size. For some purposes (eg risk analysis) it may also be useful to have a measure of the variation, such as *the variance* (see Chapter 5). In the case of a simple random sample, as described below, the variance (now written as S^2 to emphasize that it is the square of the standard deviation) is estimated by;

$$S^2 = \sum (X_i - \bar{X})^2/(n - 1)$$

where X_i is an individual observation and $\bar{X}$ is the mean.

Each of the statistics discussed above relates to a single characteristic or measurement for each household. However, the objective in farm household surveys is to arrive at a description of the whole system, which requires estimates of many inter-related characteristics. A problem then arises in deciding which statistics to use to describe the typical farm. This is often referred to as the 'modal farm', but clearly it is most unlikely that any individual farmer will fall into the modal class for every variable that is measured.

The alternative is to create a theoretical model of an imaginary farm that is typical of the sample. In taking this approach it would be inappropriate to use the sample mode for each variable, since this measure is unsuited to accounting and other arithmetical manipulations. For instance, we cannot assume that the modal quantity of maize produced times the modal price equals the modal value of maize produced. Mean values, on the other hand, can be manipulated in this way. Hence there is a stronger case for using the mean of each variable in describing and analysing the typical farm. The only possible disadvantage in using the mean is that for indivisible items such as cows or machines, unrealistic fractions may result. However, it is questionable whether this need invalidate the analysis.

Another advantage in using the mean, rather than the mode, is that we can measure its 'precision' as an estimate of the true population mean. For a simple random sample the error of the mean is calculated by

$$\text{Standard error} = \sqrt{(S^2(1 - f)/n)}$$
$$= \text{approx } \sqrt{S^2/n} \text{ when f is small}$$

where S^2 and n are as already defined and f = sampling fraction = n/N where N is the population size.

The standard error may be used either (i) to estimate a confidence interval for the population mean, such that we can assert with a given probability (eg 95 per cent) that the interval actually contains the population mean; or (ii) to test hypotheses regarding the population mean (see any basic statistics text, eg Freund, 1979). It is argued that presentation of a confidence interval is more meaningful and useful than a single point estimate of the population mean in descriptive studies, since it gives some guidance as to the precision of the estimate. Hypothesis tests may be used in on-farm testing of innovations to investigate whether there is a significant difference (one unlikely to have occurred by chance) in performance between adopters and non-adopters.

Several points should be noted, however. First, estimation of confidence intervals and hypothesis tests are only valid if appropriate random sampling techniques are used. Second, in these circumstances precision can be increased by increasing the sample size (note that the standard error is proportional to $1/\sqrt{n}$). More sophisticated sampling techniques may further increase precision for a given sample size or survey cost. Third, in practice, non-random sampling methods and measurement errors may introduce *bias* in the estimation of the population mean. The overall precision or size of the error depends upon both sampling error and bias

$$(\text{expected error})^2 = (\text{standard error})^2 + (\text{bias})^2$$

the latter often being much larger in practice.

There is probably a trade-off between these two influences. Sampling error can be reduced by increasing the sample size but, given a limited budget, this will necessitate less careful measurement on the individual farm with a possible increase in measurement bias.

Sampling

(i) *Why random sampling is desirable*

The sampling problem is to decide how to select the sample from the population. This sounds, and indeed is, a simple thing to do but unless we ensure that there is no bias involved in our sampling method, there is no hope whatever of our being able to make scientific statements about the population from the knowledge we obtain from the sample. It is by no means easy to ensure that there is no bias.

Suppose, for instance, the agricultural extension service is asked to recommend names of farmers likely to be willing to co-operate in providing farm management data. These farmers are likely to be more progressive than their neighbours and may have introduced new techniques not commonly employed on the majority of farms in the population. If this error is avoided by eliminating these farmers from consideration when selecting the sample, this

would be little better, for the bias would be in the opposite direction.

We do not usually know what biases there are in our sampling procedure if we choose it for reasons of mere convenience, speed, or cheapness, or because it has no obvious disadvantages. In sampling it is never enough not to have detected a bias; the sample should be drawn in such a way that no possibility of bias can arise. We are only really safe in this respect if the sample is selected in some way which is completely unrelated to any conceivable variable. To ensure this, we employ a chance mechanism to select the sample, that is we take a random sample. With a simple random sample every farm in the population has an equal chance of being selected.

(ii) *The simple random sample*

The random sample is therefore the ideal to be aimed at to avoid bias. However, a random sample is not always possible for farm management surveys. Thus a great deal of information, some of it of a highly personal nature, must be collected over at least one cropping season and preferably longer. This may require many visits by the enumerators and may take up a great deal of the farmer's time. It is therefore essential to find farmers who are able and willing to co-operate. Not all members of a random sample will be agreeable. Furthermore, in many parts of Africa there is no complete list of all the farmers in the population. Without such a list or 'sampling frame' it is impossible to ensure that every farmer has an equal chance of being selected.

For some purposes, such as land use surveys, it is possible to use areas of land (or their equivalent on maps) as the sampling frame, but where, as with a farm management survey, contact with the individual farm families is necessary, the best frame to use is one based on a list of the human population. Such lists may be prepared from the returns of the most recent population census, or, in their absence, from the records of local administrators, tax collectors or a centralized marketing agency. Most of these lists are likely to be either out of date, or incomplete, or both. If no comprehensive and up-to-date information for a sample frame exists, it may be desirable to make a reconnaissance survey of all farms covering only a few items, such as farm area, type of land and family size, in order to compile a complete list of farms in the area. Thus every effort should be made to obtain a complete sample frame and to select a random sample. Where this is not possible, the danger of bias must be borne in mind.

(iii) *Systematic sampling*

Systematic sampling involves choosing every jth member of the population systematically, where 1/j is the desired sampling fraction. Thus a 5 per cent or 1 in 20 sample of households in a village might be obtained by selecting every 20th dwelling passed in a tour of the village. It is generally easier to draw a systematic sample than a simple random one, but there is a danger of introducing bias if the sample units are not arranged in a random order.

(iv) *Stratification*

There are possible modifications to the simple random sample in which every farm has an equal chance of selection, although these modifications involve random selection at some point. For the 'stratified random sample', the population is divided into a number of groups or strata. These strata may consist of: (1) administrative units, (2) ecological/agricultural zones, (3) village or farm size groups, or any other means of classifying farms. Within each stratum a random sample of farms is selected, which means that every farm has an equal chance of being selected. This chance, however, might not be equal to that in a different stratum of the population. A stratified random sample is thus, in effect, a collection of simple random samples from a collection of populations.

It is generally the case that a stratified random sample gives more precise results than a simple random sample, especially if the strata are selected so that the variation between strata is as large as possible and hence the variation between farms within each stratum is minimized. The results are more precise, simply because the variation within each stratum is less than the variation in the whole population. However, in order to define the strata, it is necessary to have some additional information on the population, besides the sampling frame. This additional informa-tion will obviously be available if the sample is to be stratified by administrative units, but this method of defining strata is likely to be less effective in improving precision than stratifying by ecological zone and farm size.

Obviously, since many items are being recorded on each farm, one basis of stratification may not be equally effective in improving precision for each item. For example, the types of crops grown and the area of each crop per farm are likely to differ considerably between climatic zones *but* family sizes or the amount of capital used might vary more between farms within zones than between zones. Unless we are very fortunate, therefore, we must expect the gains from stratification to be relatively modest *but* it will practically always bring about some improvement for every item, no matter what the basis of stratification.

(v) *Cluster sampling*

The random cluster sample involves dividing the population into a number of groups. A random selection is made from these groups. All the individuals in the chosen groups then constitute a cluster sample. Whereas, with the stratified random sample, all groups or strata are included but only a sample of farms within each group are surveyed; with the random cluster sample only a sample of groups are included but all farms within the sample groups are surveyed. Unless the clusters are very carefully defined so that each one includes as much variation as possible, or reflects the full range of variation in the whole population, this method is likely to be less precise than simple random sampling for a given sample size. However, its big advantage is that it is likely to be cheaper than other forms of sampling, because the cost of enumerator's travel from one farm to another is much reduced. Hence the level of precision per unit expenditure may be increased.

Random cluster sampling is particularly useful (1) where there is no population list to serve as a sampling frame, and (2) where there is a large dispersed population or where communications are bad. Cluster sampling was used by Bessel *et al*. (1968) in Zambia.

Generally speaking, some of the advantages of both techniques can be obtained by means of a multi-stage random sample. For a two-stage sample, the population is divided into a number of groups, villages, for example: a simple random selection is made from the groups; then a simple random selection is made from the farms in each selected group. All the individuals selected in this way, taken together, constitute the two-stage sample. Thus the two-stage sample may be viewed as a cluster sample, in which only a sample of the farms within each cluster are studied, or a stratified random sample in which only a sample of the strata are included. Most of the field enquiries in the agricultural sector in developing countries have been based on multi-stage samples. Thus the first stage groupings may be ecological/agricultural zones; the second stage groupings villages; the third stage groupings farms or families; and for some purposes the fourth stage groupings are individual plots.

Where there are no population data available to serve as a sampling frame, ecological zones and villages may be distinguished and sampled from aerial photographs or maps if available. Each village in the sample may then be subjected to a population census in order to provide data for sampling farms at random within the villages.

This very brief review of sampling methods should show that selection is by no means the simple and obvious matter that it at first appears. Before embarking on any survey it is advisable to get the help of a statistician or to study the theory of sampling methods before drawing the sample.

One general point regarding sampling is worth noting, namely that it is *sample size* and *not* the fraction of the population sampled which almost entirely determines the precision of estimation for a given population. For most purposes a sample size of thirty farms in each stratum for which an independent estimate is required is probably adequate. There is little point in surveying a sample of a thousand or more farms. Resources would be better used in improving the accuracy of the data collected or in collecting additional data. Even where the number of farms studied is an insignificant fraction of the total population, a random sample of sufficient size can be used to draw reliable, unbiased results and to test the accuracy of these results. If, however, it is impossible to draw a random sample then it is important to check

as thoroughly as possible whether the results are biased in any way.

Where a survey is made in just a single year or only a few years, the years are in fact a sample from the whole population of an infinite series of years. Random sampling is not possible in this respect so it is important for the investigator to determine to what extent the information gathered each particular year represents normal or average conditions, particularly for crop yields, animal production and price levels. This, of course, does not apply where farm management surveys are made continuously year after year. Indeed, there is much to be said for establishing surveys on a permanent basis. Farm conditions and factors which influence farm business are constantly changing. Thus data rapidly become outdated. After a farm management survey has been repeated in the same area for a number of years, the data become more and more accurate, and the time involved and money spent diminish because farmers become more familiar with the nature of the survey and the type of information required. Enumerators become more experienced and do not need to repeat the initial training. Furthermore, data from repeated surveys make it possible to identify trends in yields, prices and factor inputs.

Questionnaires and schedules

There are two types of form that may be used:
  (i) the schedule for collecting factual information, in tables or lists;
 (ii) the questionnaire for collecting opinions, attitudes and aptitudes by asking the respondent questions framed in a precise way.

The schedule is often designed for ease and convenience of coding and summary of the data, although it is also necessary to set it out in such a way that the enumerator is unlikely to miss any items. Sometimes sets of schedules are bound together to form record books. One possible set of schedules for farm management data collection and analysis have been designed by FAO (Friedrich, 1977).

With a questionnaire it is important that every respondent should be asked the same question in the same way. It is therefore necessary to translate the questions into the local language on the questionnaire to avoid any slight misinterpretations by the enumerator.

All the terms used in schedules and questionnaires must be clearly understood by enumerators and agreed before the survey starts. Difficulties may arise over the definition of 'a farm' for instance. It may be defined as 'all the land and other resources under the control of one farm family', but then problems may arise in defining the 'farm family' and deciding how to treat resources under family control but not used in farming. The correct translation of local crop names must also be agreed.

Pre-testing of schedules and questionnaires is highly desirable, either as part of a pilot survey or as part of the training programme for enumerators. This allows the opportunity to correct omissions, or ambiguous questions and to discover terms, the meaning of which may not be clear to farmers or enumerators.

Organizing the survey

Preparation

The organization of a survey is a major administrative task which involves:
  (i) formulating objectives,
 (ii) delineating the study area,
(iii) choosing samples,
 (iv) designing and testing questionnaires,
  (v) selecting and training enumerators,
 (vi) preparing for their needs in the field and back-up services in the office,
(vii) carrying out a pilot survey,
all before the main survey can begin. Thus it is important that adequate time is allowed for all these preparatory tasks before the main survey period and that plans and phasing of the whole operation are worked out in advance.

It is also desirable in most cases, to hold meetings with chiefs, village councils and farmers before the main study in order to explain the aims and objectives and to enlist farmers' support and co-operation.

Some investigators have thought it necessary to provide incentives in the form of free issues of fertilizer or other inputs or in the form of cash. However, apart from the cost, the promise of a gift may alter the farmer's behaviour so that it becomes atypical. It is likely that observing local customary procedures of communication and keeping farmers informed at all times about the purpose and progress of the study is more important than the provision of financial or physical incentives.

Arrangements must also be made for housing, transport and equipment for enumerators, as well as communications for returning questionnaires, supervision and payment of wages. Generally, the enumerators can be left to make their own accommodation arrangements but it is important that they should live in the survey area to minimize travel time and cost.

Generally enumerators need some form of transport to visit farms and this can prove a costly item. If cluster sampling or multi-stage sampling is used, it may be convenient and not too costly to take a small group of enumerators by motor vehicle to a sample village, dropping them one by one at sample farms or allowing them to walk between farms. Where the sample farms are too widely scattered for this approach, it may be necessary to provide each enumerator with a bicycle or, where distances are greater still, a motorcycle. Careful planning and budgeting is needed to find the most suitable form of transport in terms of convenience and cost.

Enumerators require, besides a stock of schedules and questionnaires, clipboards, and writing materials. They may require other equipment, depending upon the records to be collected, such as surveying equipment for measuring areas of plots of land, harvesting tools and weighing balances for crop-cutting and weighing of yields or stop watches for timing labour use. All such equipment should be acquired in advance, before the main study begins.

Communication between the enumerator and the survey office is probably best maintained by regular supervisory visits, when the enumerator can be paid, completed survey forms can be checked and collected while progress and problems can be discussed. Unless enumerators are very experienced and trustworthy employees, regular supervision is essential.

The enumerators

The personality and behaviour of the enumerators has an important effect on the willingness of farmers to co-operate. A good working relationship must be established. Thus choice of enumerators, their training, motivation and supervision are all important considerations.

Enumerators must be fluent in the language used by the farmers and it is desirable that they should know something of local farm conditions and practices so that they ask questions intelligently and check on the accuracy of the farmer's replies.

There are, therefore, advantages in recruiting local inhabitants of the survey area. However there are also possible disadvantages if the enumerator is a member of a particular faction, religious group or political party whose opponents may refuse to co-operate. Also it may be difficult to sack an enumerator who is unsatisfactory in the work, if he is a member of the local community, since this may turn farmers against the study and create problems for his replacement.

Another consideration in choosing enumerators is the educational standard required. This must depend upon local circumstances. In some places there may be unemployed university graduates who could be recruited for such work whereas in other places, primary school leavers are the most highly educated people one could hope to recruit. Generally speaking, a high educational standard is not needed provided that the applicants are reasonably literate and numerate and adequate training is provided. Selection may be based on an interview and a simple test of ability to write clearly and make simple calculations.

The possibility of employing part-time enumerators should be borne in mind. People such as extension agents or school teachers may be used. However, there is always a problem of dual allegiance which makes supervision and control difficult. There is a danger that they will withdraw from the project when an opportunity for promotion occurs or when annual leave is due. University students may be used if the main survey work can be restricted to the vacations. Such experience can be very valuable to students of agricultural subjects.

Motivation of enumerators is important and they should be paid adequate wages comparable with those they could earn in similar employment elsewhere. Ideally, where there is regular and fairly continuous survey work in progress, a permanent cadre of professional enumerators should be established with opportunities for promotion resulting from good service. However, this may not be possible if there is inadequate work to keep them fully employed.

Whatever the background of the enumerators, some training is needed before they start work in the field. Generally a period of two to three weeks, made up of say one week of office training and the rest in field training, will be adequate. During the office training the purpose and importance of the study can be explained. The survey questionnaires and schedules should be studied in detail with some discussion of the ways in which the results will be summarized in order to give trainees a thorough understanding of their interview procedures. They should also be instructed in the techniques of assessing areas, weights and measures. Field training is devoted, in the main, to giving enumerators practice in completing questionnaires and schedules with farmers.

The number of farmers each enumerator can be expected to visit each week must depend upon

(i) the time it takes to travel from one farm to another, which in turn depends upon distances and means of transport,
(ii) the time it takes to complete each interview which depends upon the amount of information collected and the method of measurement used,
(iii) whether farmers are only available at certain hours for interview or at any time.

A decision may, perhaps, be delayed until after a pilot survey which will give a clearer picture of what is possible but as a crude guide, four or five visits per day or twenty to twenty-five visits per week should be possible if sample farms are relatively close together. When most of the time is spent in travelling the number that can be visited is, of course, reduced.

Frequency of visiting farmers

A critical decision which affects both the cost per farm surveyed and the accuracy of the data collected, is the number of times each of the chosen farms is visited. It may range from once only to daily visiting over a whole year or longer. There is apparently a trade-off between savings in cost and gains in accuracy per farm. However, certain gains in reliability are obtained by increasing the sample size so if the reduction in cost per farm allows an increase in the number of farms studied there may be an overall *gain* in reliability of the results.

In part the decision may be whether to rely on recall (ie the farmer's ability to remember inputs used and yields obtained in the past) or direct observation. Clearly direct observation of amount of seed used as well as amount of crop harvested is impossible when the farm is only visited once. However, even with quite frequent visiting, it is necessary to rely on the farmer's recall, though only over a short period since the last visit. Accuracy is likely to be greater, the shorter is the period of recall.

The scope for saving by infrequent visiting depends upon the complexity of the farming system. In the case of a simple system with a single, short cropping season, no perennial crops or livestock, a single visit just after harvest might be sufficient to provide acceptable data. More frequent visiting would probably be essential to study systems with two or more cropping seasons, some perennial crops and livestock.

A distinction may be made between (i) 'single-point data' such as area of land, numbers of livestock, or productive trees and stocks of machines, equipment and materials and (ii) continuous data such as daily labour use and quantities of other inputs and outputs. Whereas 'single-point data' may be collected in a single visit, reliable records of continuous data may require regular and frequent visiting.

Within each of these categories of 'single-point data' and continuous data, a further distinction may be made between 'registered' and 'non-registered items'. The former consist of items such as rented land areas, hired labour use or cash crop sales which are associated with market transactions and therefore are 'registered' in the farmer's mind if not on paper. Non-registered items include family labour use and household consumption of food-stuffs which are far less likely to be recorded. Registered items can be recalled more easily and hence can be collected

satisfactorily with infrequent visiting. Overall then, reliable information on single-point, registered items may be collected in a single visit, but to get accurate information on continuous, non-registered items may require regular and fairly frequent visiting; say every two or three days. (See Collinson, 1979 op. cit.)

Where farmers have more than one dwelling, for instance, where, as in parts of central and southern Africa, the cattle post is located at quite a long distance from the cultivated plots, it may be necessary to visit each of the holdings to make observations and collect records. The risk that the farmer may not be 'at home' on a single visit is perhaps greater than in a more settled system of farming.

Measurement

Measurement of land areas

The area of land farmed is clearly a single-point item but it may not be registered; that is the farmer may not have a very precise idea of the exact area. Direct measurement may be necessary.

The first requirement is to locate and identify which plots or fields are cultivated by the sample farm household, since many family farms are made up of several scattered plots. Omissions and errors may occur at this stage for several reasons.

  (i) The farmer may not wish to disclose how much land he controls because he fears he will be taxed upon it or for other reasons.

 (ii) Wives or other household members may have their own plots which the family head may fail to mention although strictly speaking these plots form a part of the family farm.

(iii) The farmer may only mention those to which he has long-term usufructory rights and may fail to mention land which is rented or pledged.

(iv) He may fail to mention very distant plots.

Having identified the plots on the ground it may be useful to make a sketch map of the whole farm for inclusion with the other records as a visual check that information is collected on all the plots. It is also desirable to paint some identifying mark or number for each plot, on a convenient tree or rock.

Difficulties may arise in defining crop boundaries, especially where crop plants tend to spread or

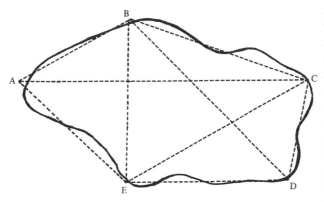

Fig. 11.1 Area measurement by triangulation. The sum of the areas of triangles ABC, ACE and CDE may be compared with the sum of the areas of triangles ABE, BDE and BCD for checking purposes

ramble. Furthermore boundaries may change over the season as more land is cleared or some reverts to bush. More than one visit will be necessary to discover this.

Generally, if the farmer does not know the area of his plots, direct measurement is required. Methods which might be used include

(i) triangulation (ie dividing the area up into triangles) and measuring the sides by pacing, surveyors chains, tapes or a measuring wheel (see Figure 11.1),
(ii) measuring offsets, perpendicular to a straight, base-line using survey chains and tapes,
(iii) compass survey, again using any of the devices mentioned above to measure distance,
(iv) plane table,
(v) aerial survey, though costs may be prohibitive for this last method.

Some of these alternative methods are discussed in Upton & Anthonio (1965) Appendix II and Hoyoux (1979).

Areas are either calculated using the formula for the area of a triangle for instance, or estimated from a scale drawing over which a squared grid, of the same scale, is placed. The area is then found by counting the squares.

Inter-cropping raises special measurement problems. The simplest approach, and perhaps the most realistic is to treat a mixture of, say, maize

inter-cropped with beans as a single crop different from sole crop maize and from sole crop beans, with its own pattern of labour requirements, costs and returns. Unfortunately, mixtures frequently include many more than two crops and since the proportions in the mixture can vary, the range of possible alternative combinations is practically infinite. Hence, in order to distinguish between different crop mixtures (and possibly to assess their relative merits), some information on plant densities is needed. It may possibly be based on visual assessment of the plot by the enumerator, or on quantities of seed used or on plant counts of sample areas within the plot. None of these methods is wholly satisfactory.

Special problems arise in assessing the areas of fallow land and communal grazing land per family. However, in both these cases, the collection of accurate data may not be considered very important. The area of fallow land might be estimated by asking the farmer how many years of fallow and how many years of cropping occur in a rotation, then multiplying the area cropped by the ratio years of fallow/years of cropping.

This is not very reliable, especially where different rotations are practised on different plots or where the length of fallows is changing over time.

For communal grazing land the only solution may be to estimate the total area and divide it by the number of families using the land.

Measurement of labour inputs

A very crude assessment of the total labour input can be based on the numbers of men, women and children in the labour force multiplied by the number of hours each is expected to work. However, since the number of hours worked can vary widely from one individual to another, the margins of error may be very large and this method gives no detail of the allocation of labour between different activities.

For most farming-systems analysis and farm planning, information is needed on the seasonal pattern of labour requirements for individual crops (or crop mixtures) and livestock enterprises. As already mentioned, labour use is continuous and (except perhaps for hired labour) unregistered, so the collection of reliable data requires regular and frequent visiting.

At each visit the enumerator records the date, and (day by day since the previous visit) the operations carried out on each crop plot and the time spent on each. Similar details are collected for work connected with livestock.

For completeness, as a means of checking records and for other uses, it is desirable to record hours of sickness, hours spent in entertainment and relaxation and hours of non-farm work. Difficulties may arise in defining whether a particular task is farm or non-farm work, for instance processing and marketing of produce. Decisions on categories of work must be made and agreed by all enumerators before the survey begins. Time spent travelling to and from the fields may take up a significant portion of the working day. It is normal practice to include travelling time as a part of the work time.

In collecting labour records it is necessary to separate different categories of worker, say (i) family head (ii) other adult male family members (iii) adult female family members (iv) children (under 14 years old) of the family (v) hired men (vi) hired women (vii) hired children. This is necessary because there is generally some division of labour between sexes and classes of labour so they are not perfect substitutes. Even when one category can substitute for another, hourly work performance may vary with physical strength and motivation. Thus, whilst on light work there may be little difference in performance between men and women, on heavy bush clearing and cultivations men may achieve much more per hour. For hired labour, wage and other payments must be recorded together with information on associated bullock or equipment hire.

Assessment of hours worked may be unreliable without clocks and watches. It may be necessary to relate periods of work to the movement of the sun, or to prayer or meal-times in questioning the farmer.

All this assumes that labour records are based on recall by the farmer of the hours worked. However, direct measurement of rates of working, using work-study techniques may be an alternative, (see Farrington, 1975). The time spent carrying out a specific task on a measured area of land or quantity of produce, is timed accurately by stopwatch. The advantages of this approach are

(i) work-study requires a far smaller volume of labour data, than do frequent visit surveys to produce mean values with comparable errors;
(ii) the costs involved in the separate surveys of areas and yields required for estimates of per hectare labour requirements by frequent-visit surveys are avoided by work-study where measurement of the work achievement is performed directly at the end of each observation;
(iii) the directness of the technique excludes the possibility of respondent confusion or omission inherent in memory-based techniques.

The disadvantages are

(i) there are certain operations for which it is practically impossible to measure the work achieved during observations of only a few hours' length, eg tobacco curing or bird scaring.
(ii) work-study only provides information on work rates, survey data are still needed to provide information on the seasonal pattern of operations and the number of times they are carried out.

Nevertheless, some saving might be made by using a combination of survey and work-study.

Most farm-survey data collectors in the past have been concerned to find a means of aggregating different categories of labour into a total labour input in 'standard man–hours' or 'man–equivalents'. Weighting factors are used for converting the work of women and children into man–equivalents; for instance, weights of 1.00 for adult males, 0.67 for adult females and 0.33 for children under 14 have been proposed for this purpose. However, for reasons given above any such weighting system must be arbitrary and there may be advantages in keeping labour records subdivided into separate categories.

Measurement of crop yields

Very often harvesting is fairly continuous, rather than a single-point operation, and unless the crop is sold immediately quantities are not registered. Thus estimates based on long periods of recall are likely to be vague and inaccurate. Regular visiting is desirable over the harvest period so that amounts harvested can be recalled more easily.

To avoid total reliance on recall, direct

measurement by crop-cutting on sample plots may be used. These sample plots should be marked out within the standing crop sometime between planting and harvest, generally the earlier the better as this limits crop damage. Each sample plot is of a standard area (eg 3 metres square or 9 sq metres) marked out with pegs and wire or string, but is located randomly within the whole cropped area. The number of samples taken in any one parcel of land ranges from one up to ten or more but it must depend, in part, on (i) the size of the parcel (ii) the variability of the crop stand, (iii) the level of accuracy desired, and (iv) the costs that can be afforded (see Spencer, 1972).

The sample plots are cultivated along with the rest of the field but are harvested separately, the yield from each plot being weighed accurately. Since the weight of most crops can vary significantly according to their moisture content, it is advisable to measure the moisture content when weighing the plot yield so that the yield can be adjusted to a standard moisture level. The yield estimates obtained are then multiplied by the total area of the crop to arrive at an estimate of total output. The main disadvantages of crop cutting are

(i) it is somewhat inconvenient for the farmer so he may not be ready to co-operate;

(ii) it is costly and time-consuming for the enumerator, especially where many sample plots are involved;

(iii) yields are usually over-estimated because the useful yield (actually available to the farmer) is often less than the total biological yield which is measured from the sample plots. (See Zarcovich, 1965).

(iv) it may be difficult to arrange the crop cutting at the most appropriate time, when the rest of the crop is being harvested, especially where mixed cropping is practised and the component crops are harvested at different times.

It has been suggested that experienced enumerators may be able to make reasonably accurate estimates of crop yields simply by looking at the mature crop and judging the yield. Clearly, this must give rather crude estimates, less satisfactory than actual measurement.

For some tree crops, where the fruit grows in bunches yield estimates can be based on a count of the total number of bunches and sample weighings of a few of them.

It is a good idea to ask farmers at some stage whether they consider the yields obtained this year to be about average, better or worse than average to give some idea as to whether the results are typical.

Other yields and sales

For livestock such as dairy cows or laying hens, yield recording, if it is not already done by the farmer requires regular visiting by the enumerator. Births, deaths and slaughterings of most classes of livestock are more easily recalled and can be collected at relatively infrequent intervals.

If records are kept of produce disposals, both sales and home consumption, these may provide a cross check on the estimated yields and total production. Discrepancies may arise as a result of wastage, losses in store, gifts and so on.

Another reason for recording sales is to collect data on the market prices obtained. In order to carry out a financial analysis of the farm business, total gross output of the various different farm products, is evaluated in money terms, using current market prices. Hence price data are an essential part of a farm business survey. Where some produce is marketed through a co-operative or a marketing board whilst other produce is sold in local markets, it is useful to record this too.

Measurement of capital assets

On practically every farm there will be certain capital assets which must be taken into account in farm business analysis. These may include livestock, standing crops, irrigation works, drainage and other land improvements, buildings, machinery and equipment as well as stocks of food, seed and agricultural chemicals both purchases and home produced. Increases in value of certain assets such as growing livestock and tree crops or stocks of food and seed represent a part of the total farm gross output, whereas decreases in value (depreciation) of machinery and equipment represent costs of production.

Generally information on a farmer's capital assets can be collected in a single visit or preferably two visits, one at the beginning of the production period (opening valuation) and one at the end (closing valuation).

The first task in assessing capital assets is to make a list or inventory. This should be fairly straightforward except possibly for recording the numbers of free-ranging livestock or the quantities of grain and other produce on hand.

Valuation of capital assets can raise problems. For items which are commonly bought and sold, such as stocks of food, seed and chemicals, livestock and some tools and equipment the current market prices can be used, *but* where there is no established secondhand market, as is probably the case for permanent crops, irrigation works, other land improvements and some kinds of machinery and equipment, this is not possible. In theory, the present value of such assets should be based on estimates of their future productivity, but, since such estimates would be largely guesswork, the normal practice is to take the original purchase price or cost of establishment and subtract a depreciation allowance for the age of the asset. This is not entirely satisfactory since prices and costs may change over time and the estimation of depreciation rates is rather arbitrary. It is therefore advisable to use standardized average prices, costs and depreciation rates on all the survey farms when valuing capital assets.

It may be desirable to collect information on the farmer's cash assets, his credit and his indebtedness but farmers may be reluctant to provide such 'sensitive' information unless there is very good rapport between enumerator and farmer. However, such information although valuable and interesting is not essential for analysis of the farm business. If it is to be collected the following suggestions should be borne in mind.

(i) Such information is best collected towards the end of field work.
(ii) Questionnaires on those items should be short and simple.
(iii) It is better to interview the farmer in private (Spencer, 1972).

Measurement of other inputs and expenditures

Although stocks of seeds, fertilizers and other agricultural chemicals may be included in the capital valuations it is necessary to record their use and levels of application for purposes of farm business analysis. Where such inputs are purchased their source and price should be recorded. Similar considerations apply to livestock feeds and medicines.

In order to assess the inputs used on individual enterprises, detailed recording is needed. Local measures, such as bowls or even handfuls may be used in distributing seed, fertilizer or chemicals while livestock feeds may be measured in bundles for instance. Average weights must be estimated, by sample weighing, for all these local measures to convert the quantities into more widely recognized units.

Records of hours worked by oxen, power tillers or tractors, irrigation pumps and other equipment may be desirable for farm planning purposes but are not essential for farm business analysis. However, purchased inputs of spares and materials such as fuel or lubricating oil must be recorded.

Information on other sources of income, household expenditure and food consumption is valuable as a cross check on other information collected, besides being interesting and useful in itself. However, such information is not necessary for analysis of the farm business and may be costly and difficult to collect. A decision must be reached before the survey begins, whether the advantages of having these data outweigh the additional costs.

Further reading

Ancey, G. (1975). *Niveaux de decision et functions objectif en milieu Africain*, Paris, INSEE, AMIRA Note de Travail No. 3

Bessell, J. E., Roberts, R. A. J. & Vanzetti, N. (1968). *Survey Field Work,* Universities of Nottingham and Zambia, Agricultural Labour Productivity Investigation. (UNZALPI) Report No. 1

Byerlee, D., Collinson, M. P., Perrin, R. Winkelmann, D., Biggs, S., Mozcardi, E., Martinez, J. C., Harrington, L. & Benjamin, A. (1980). *Planning Technologies Appropriate to Farmers: Concepts and Procedures*, El Batan, Mexico, CIMMYT

Casley, D. J. & Lury, D. A. (1981). *Data Collection in Developing Countries*, Oxford, Clarendon Press

Clayton, E. S. (1961) Economic and Technical Optima in Peasant Agriculture, *Journal of Agricultural Economics*, **14**(3) 337

Collinson, M. P. (1982) *Farming Systems Research in Eastern Africa: The Experience of CIMMYT and Some National Agricultural Research Services 1976–1981*. East Lansing, Michigan, Michigan State University, Dept. of Agric Econs Development Paper No. 3

Farrington, J. (1975). *Farm Surveys in Malawi*, University of Reading, Department of Agricultural Economics and Management, Development Study No. 16

Freund, J. E. (1979). *Modern Elementary Statistics*, 5th edn London, Prentice-Hall

Friedrich, K. H. (1977). *Farm Management Data Collection and Analysis*, Rome, FAO

Gastellu, J. M. (1980). Mais où sont donc ces unités économiques que nos amis cherchent tant en Afrique? *Cahiers ORSTOM, Series Sciences Humaines*, **27** (1–2)3

Hoyoux, J. M. (1979). *A Manual: Measuring Size of Small Farms*, Ibadan, IITA, Discussion Paper 6/79

Kearl, B. (ed) (1976). *Field Data Collection in the Social Sciences: Experiences in Africa and the Middle East*, New York, Agricultural Development Council

MacArthur, J. D. (1968). The economic study of African small farms: some Kenya experiences. *Journal of Agricultural Economics,* **19**(2) p. 193

Norman, D. W. (1973). *Methodology and Problems of Farm Management Investigations: Experiences from Northern Nigeria*, African Rural Employment Paper No. 8, Department of Agricultural Economics, East Lansing, Michigan State University

Spencer, D. S. C. (1972). *Micro-level Farm Management and Production Economics Research among Traditional African Farmers: Lessons from Sierra Leone*, African Rural Employment Paper No. 3, Department of Agricultural Economics, East Lansing, Michigan State University

Stuart, A. (1964). *Basic Ideas of Scientific Sampling*, London, Griffin

Upton, M. & Anthonio, Q. B. O. (1965). *Farming as a Business*, Oxford University Press

Yates, F. (1981). *Sampling Methods for Censuses and Surveys*, 4th edn, London, Griffin

Zarcovich, S. S. (1965). *Sampling methods and censuses*, Rome, FAO

12

Analysis

Coding and processing data

At an early stage in the planning of a survey a decision should be reached on how the results are to be analysed in terms of both (i) the types of analysis that will be made, and (ii) the data handling methods to be used, including whether or not to use a computer.

Whatever methods of analysis are to be used, data coding is recommended. For quantitative, numerical information this simply means setting out the figures collected on the farm in a convenient layout for further summary and analysis. In the case of qualitative data such as sex of family head, soil type, or statements of opinion, coding consists of allocating numbers to each of the alternative possible answers and using these numbers in further analysis rather than the written answers (eg a male head might be coded as 1 and a female head as 2). The reason is simply that it is quicker and more convenient to manipulate numbers rather than written answers. Coding tables may be incorporated in schedules and questionnaires or enumerators may be required to transfer their records to coding sheets daily when they return from field work.

Use of a computer must depend upon whether computing facilities are available. Even where these facilities are available it is by no means certain that the use of a computer is justified. Against the advantages of high-speed calculations must be set the costs not only of the use of computer facilities but also of learning how to prepare data for entry to the computer and how to write instructions regarding the analyses to be carried out. These preparatory stages can be very time-consuming. Many software packages are now available, on both micro- and main-frame computers. Of particular relevance is the 'Farm Analysis Package' (FARMAP) developed at FAO (FAO, 1983) and the Statistical Package for the Social Sciences (SPSS) (Nie *et al.*, 1975) which is designed for the analysis of farm and other surveys. Although the former is specifically designed for farm survey analysis the latter may be more attractive because of its greater flexibility if additional analysis including statistical calculations are intended.

Whether all the analysis is carried out by hand with pocket calculators, or whether a computer is used, analysis and summary of results together with the writing up, are very time consuming. Experience suggests that these final stages take at least as long as the survey itself. Frequently inadequate resources of time and funds are allowed. The number and type of staff needed depend upon the data-handling methods chosen. Hand analysis requires clerical and calculating assistants while use of the computer requires computer operators for data entry and analysis.

Initial cross-tabulations

There are two broad categories of information that may be obtained from a farm study: simple variables and composite variables. The first category includes those items which are recorded directly in the field. It clearly includes farmers' statements about their attitudes and objectives but may apply to estimates of total land area, labour use or farm income. The other category refers to those items which are the result of certain calculations applied to the basic data, for instance, when separate labour records are aggregated to arrive at total labour input or accounting methods are used to estimate household income. We deal first with methods of summary and presentation of survey data in an informative way. The same

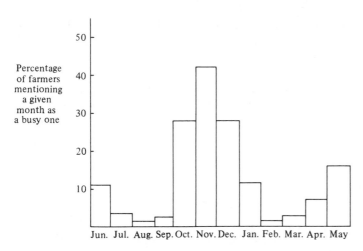

Fig. 12.1 Distribution of busy months

principles may be applied to the presentation of composite variables, such as household income per capita, once they are estimated. Accounting methods used to estimate such variables are dealt with later in the chapter. The following discussion on the summary of survey statistics is, of course, irrelevant when a case study approach is used. None the less, accounting methods are needed to analyse the case study system.

Coded responses for any single variable can be grouped into classes and represented as a frequency distribution of the number of observations or households in each class. For binary data there are, of course, only two classes, while for continuous data it is necessary to first identify the classes by defining their limits. In defining class limits the following rules are usually observed: (i) the entire range of values is divided into, generally fewer than, 15 classes, (ii) the classes are mutually exclusive so that each observation can appear in one and one class only, (iii) where possible the classes cover equal ranges of values although this may conflict with the last rule, and (iv) the number of observations in each class should be sufficient to justify showing that class separately.

It is a useful first step, in the analysis of any data item, to determine the frequency distribution of the responses. This is not too onerous a task when analysing data by hand, and is very easily achieved with a survey analysis package such as SPSS on a computer.

Having determined the frequencies it is but a simple step to express the results as relative frequencies, or percentages of the total. This, in effect, gives the probability distribution of the responses. For ease of interpretation it may be useful to plot the frequency histogram as in Figure 12.1.

A tabulation or plot of the frequency distribution of observations is useful in several ways

(i)   in data checking and validation since values lying outside the feasible range are readily identified,

(ii) in examining the form of the distribution, its spread and whether it has more than one modal class or peak (the latter case might suggest that the data come from more than one distinct population),

(iii) in identifying the modal class.

Figure 12.1 shows the frequency distribution of responses, of a sample of Zambian maize growers, as to which is the busiest month. The approach of simply asking farmers to identify critical constraints in this way is clearly cheaper than using detailed records and accounts of the whole farm system for the same purpose as described below.

A useful additional step is to prepare 'cross-tabulations' which present results for two (or sometimes more) variables at once. Each column of such a table represents the frequency distribution for variable A, *within* each class according to variable B. The

Table 12.1. *Percentage distribution of decision-makers on various activities connected with food crop production*

Activities	Decision-makers				
	Wife	Husband	Both	N.A.*	Total
Which crop to grow	44.0	45.8	9.3	0.9	100.0
Acquisition of land	12.0	85.6	2.3	—	100.0
When to plant crops	59.3	33.8	6.0	0.9	100.0
Use of fertilizers	1.9	23.1	1.4	73.7	100.0
Increase farm size	40.3	55.6	4.2	—	100.0
Leave land under fallow	44.4	50.9	3.2	1.4	100.0
Which crops to sell	56.5	35.2	6.0	2.4	100.0
When to sell crops	55.6	39.4	2.8	2.4	100.0
To whom to sell crops	44.9	46.8	5.1	3.3	100.0
Purchase of farm tools	7.4	86.6	5.6	0.5	100.0

* N.A. = No answer.

column totals then represent the total class frequencies for variable B, while the row totals represent the total class frequencies for variable A. Individual cell frequencies may be represented as percentages of either the column or the row total. The advantages of cross-tabulation over and above the benefits of examining frequency distributions already mentioned is that it may assist in identifying associations between variables.

An example is given in Table 12.1 comparing decision-making responsibilities (variable A) of different family members (variable B). From this table it appears that, in the study area of East Cameroon, male farmers or husbands are dominant in decisions regarding acquisition of land or tools or when to leave land fallow. Wives, on the other hand, are mainly responsible for deciding when to plant, what crops to sell and when to sell them. In such circumstances the

chi-square statistic may be used to test whether the apparent association between the variables (in this case the difference between the sexes) is statistically significant meaning it is unlikely to be a chance effect (for more details see Freund, 1979).

Where the objective is on-farm testing of new technology and a comparison of innovators with a control group, then the binary variable, adopter/non-adopter might be used as one of the variables to classify the data. The cross-tabulation by another variable, say farm income, would allow comparison of the two groups.

Analysis of the farming system

The cross-tabulation of responses and interpretation of the distributions is generally the only analysis that is needed for data on attitudes and objectives collected by interview. For most other variables used in describing a farming system

(a) some analysis or manipulation of the data is needed within each farm household to arrive at the desired measures;

(b) the results obtained from a sample of households are summarized by estimating the mean and perhaps the variance (see Chapter 11).

It is desirable to carry out these operations in this order; that is to analyse the system for each of the sample households before summarizing by calculating the means. This is the only way in which variation between farms can be assessed. For instance, the mean yield of maize could be estimated by dividing the total output of maize from all farms in the sample by the total area of maize; but this would provide no measure of the variation in yields between farms. Furthermore, it would provide a measure of mean yield averaged over all hectares, whereas we are here concerned to estimate the mean yield averaged over households. Only in this way can we be sure that the mean quantity of maize produced per household is equal to the mean quantity sold plus the mean quantity consumed, stored or wasted. In short, every item in the analysis must be averaged across households for the accounts of the average household to be internally consistent.

Estimates of the variance between households may be used to establish confidence intervals for the main variables measured. However, as we have seen, such estimates are themselves unreliable if the sample was not drawn randomly and the results might be biased. The other main use of variance measures is in risk analysis. Unfortunately, the variance between farm households within one season may be a very poor and unrepresentative measure of the variance between seasons. The latter is likely to be the main concern of the risk-averse farmer. None the less, having collected data from a sample of farms, the estimation of the variance is relatively easy and is probably justified.

Having considered the presentation of summary data we turn now to the logically prior question of analysis of the individual farming system. The types of data that are needed were discussed in the previous chapter. Apart from the assessment of farmers' attitudes and objectives, these consist broadly of (i) descriptive data on the farming system; the resource base, cropping patterns and livestock numbers and (ii) measures of inputs, outputs, costs and returns. Variables in this last group may be estimated for individual plots of land, for specific enterprises and activities or for the whole household.

Descriptive data

The description of available resources and the combination of crop, livestock and off-farm activities may be set out in a series of tables under the headings of land, labour and capital.

(i) *Land* There are three different ways in which the total area of land under the control of the farm family may be analysed: a) by land-use category, b) by tenure and c) by crops grown. The different land-use categories may include rainfed arable, irrigated arable, permanent crops, permanent pasture or rangeland and fallow. Further distinctions may be drawn according to soil type or topography. The sum of the areas in all these categories should equal the total farm area.

Categorization by tenure involves separation of common land from that held by individual family members. Land which is pledged or rented is separated from land which is owned or held under customary tenure. It is useful to supplement this

with information on the relative ease of acquiring additional land.

Finally the pattern of land use should be detailed in terms of the areas of different crops grown. Where there is only one crop season per year and sole-cropping is practised, description is straightforward. The sum of the areas of individual crops plus grass-land and fallow should add up to the total area of land available to the household. Where two or more crops can be grown sequentially within a year, the area of each crop should be recorded. The total area of crops (and fallow) then exceeds the total farm area. The ratio of these two totals may be calculated, as a measure of the intensity of land use (see Chapter 6).

More serious problems arise in dealing with crop mixtures, when it is difficult to assess whether the component crops are competitive or supplementary. Judgement is needed in deciding whether to treat each component crop as covering the whole area or to assume each crop covers a fraction of the area. In some cases, especially for complicated mixtures of many different crops, it may be most appropriate to treat each mixture as a separate and distinct crop. None of these methods is wholly satisfactory, and special methods of assessment based on relative crop cover may be needed for detailed analysis of mixed cropping.

(ii) *Labour* The basic regular labour force is usually made up of family members. Even hired labourers frequently live in as members of the household. Hence an analysis of the household composition, may give an assessment of the regular labour force available. Household composition is analysed by age and sex categories. Conversion factors may then be used to estimate the total labour force in standard adult male equivalents. Due account must be taken of other off-farm commitments in calculating the residual labour available for work on the farm. Household composition data may also be used to estimate total food consumption requirements. The data should be supplemented by information on the ease of hiring more labour, and the normal wage rates.

(iii) *Capital* Capital invested in permanent crops is recorded under the cropping pattern. The remaining capital items to be mentioned now are livestock and physical assets of machinery and equipment. Livestock numbers are obviously separated by species and sometimes by age and sex categories, to give flock or herd structures. For purposes of aggregation, livestock unit conversion factors may be used to arrive at a) total livestock units of each species, b) total grazing livestock units for ruminant cattle, sheep and goats, or c) grand total of all livestock units owned by the household.

Separate records may be presented for individual items of machinery and equipment used on the farm or elsewhere. However, it may be thought desirable to estimate the total value of capital assets in money terms. The total value of physical assets plus permanent crops, livestock, stored products and cash in hand minus any outstanding debts gives a measure of the farmer's 'net worth'. Given that some assets are rarely bought and sold so that estimating their value is essentially arbitrary, and given that farmers are often unwilling to disclose their financial position, the measurement of net worth may prove difficult in practice. In any case, the measure is of limited value to a semi-subsistence farmer except as a guide to his creditworthiness.

Input–output data

The objective here is to calculate the quantities of inputs used and of outputs produced per hectare of each crop or per head of each class of livestock. Crop input–output data may be estimated from individual plot records, while, for permanent crops such as oil-palms, it might be appropriate to calculate the amounts per tree.

Let us consider the measurement of inputs. Some, such as agricultural chemicals or tractor services, may be purchased or hired while others such as family labour are supplied from household resources. Inputs of both kinds should be recorded for each plot or enterprise and converted to a per hectare or head of livestock basis.

An alternative distinction may be drawn between stock and flow resource inputs. Resources which are available in the form of stocks such as seeds, fertilizers and other chemicals or concentrate feeds for livestock, can be stored. If they are not used at a

particular point in time, they can be kept for future use. Hence, it is generally not necessary to record the timing of stock resource inputs. Such inputs are generally associated with variable costs.

Resources such as regular labour, or draught animals and equipment provide a continuous flow of man–hours or oxen–hours which cannot be stored for future use in the way that seeds can. Unused labour in January will not add to the labour supply in August. The cost of the flow is fixed and unavoidable, whether the labour is actually used at a particular time of the year or not. If such resources are likely to be limiting constraints it is highly desirable that the seasonal distribution of inputs should be estimated.

Labour inputs may be recorded separately, not only for different dates, but also by age and sex of the worker, by plot or livestock enterprise and by operation. Some aggregation may be desirable in order to present the seasonal labour profile. Labour inputs for different age and sex groups may be aggregated by converting them all to standard man–days. If the labour profile is to be based on monthly intervals, then labour inputs on different dates and for different operations within the month may be aggregated to give total monthly labour input. Finally, labour inputs on different plots of the same crop may be aggregated to give the total monthly input to that enterprise. The ultimate objective is to determine the seasonal profile of labour inputs per hectare of each crop and per head of each class of livestock. Similar profiles of inputs may be calculated for draft animals or machines such as tractors.

In measuring the output of each enterprise (or plot) it is important to include both marketed and home-consumed produce. Where yields have been recorded directly the problem does not arise but where yield data are not available, then they must be estimated by combining quantities sold with quantities used in the household. Furthermore, in the case of livestock, and possibly some crops, there may be a change in the quantity on hand between the start and end of the year; and some may have been purchased or received as gifts. These items must all be taken into account in estimating the total yield. Losses due to animal mortality or crop wastage are generally excluded from the output measure.

For illustration, the total number of goats

Table 12.2. *Estimating total number of goats produced*

	Number of goats
Number sold	3
Number consumed (or given away)	2
Number on hand at end of year	5
Total I	10
Number on hand at start of year	4
Number purchased (or received as gifts)	2
Total II	6
Number produced (Total I) − (Total II)	4

produced in a household flock in one year is estimated in Table 12.2. In practice, it might be more useful to separate different age and sex cohorts (see Chapter 8). Transfers into, and out of, different age classes would then have to be taken into account.

Having estimated the total output or yield it is normally expressed on a per hectare basis for crops and per head of livestock.

Farm business analysis

For accounting purposes, in order to compare returns from different enterprises with their costs of production we need a common unit of value. Nutritional measures such as grain equivalents (see Clark & Haswell, 1964) or megajoules (MJ) of energy (see Bayliss–Smith, 1982) have been used, but, clearly, there are difficulties in evaluating non-food items such as cotton or rubber in this way. Money values, on the other hand, can usually be estimated for all commodities including those produced mainly for subsistence. In most African situations, subsistence crops surplus to household requirements are sold, and the prices received may be recorded.

With data on yields and prices for each enterprise we can calculate the enterprise gross output as follows:

$$\text{gross output} = \text{yield} \times \text{price}$$

Where there is more than one product such as grain and straw, or calves and milk, the total gross output is

the sum of the values of the joint products. Also, if permanent crops or livestock change in value between the start and end of the year, the gross output measure must be adjusted accordingly. Crop gross outputs are usually expressed on a per hectare basis or for some permanent crops per tree. Livestock gross outputs are expressed per head or per livestock unit. The whole farm gross output is simply the sum of the gross outputs obtained from the individual enterprises.

Costs of production are usually separated into (a) variable or direct costs and (b) fixed costs or overheads. In earlier chapters on the theory of production we assumed that any input may be varied; the distinction between fixed and variable costs then depends upon which inputs are assumed to vary. However, in farm business analysis, it is convenient to standardize the classification in the following way.

Variable costs	Fixed costs
Crops	
Seeds	Rent or costs of land use;
Fertilizers	wages or costs of labour use;
Sprays	interest on capital invested; depreciation of machinery,
Livestock	draft animals, equipment
Livestock feeds	and buildings; maintenance
Veterinary medicines	and repairs

There is no general agreement regarding machinery fuel and running costs or temporary hired labour. Although machinery running costs clearly do vary with the amount of use, it is convenient to treat them as fixed costs for general farm business analysis. Temporary hired labour is also a variable cost but if only a few farms employ casual workers it may be more appropriate to treat the labour costs as being fixed on all farms. However, where casual hiring is normal practice, say for cotton harvesting, then the cost may be treated as variable.

The distinction between variable and fixed costs has traditionally been drawn on the basis of the difficulty of allocating fixed costs to individual enterprises. However, in African agriculture, the distinction might be based on the difficulty of evaluating the fixed costs. It may be noted that the variable costs generally correspond with stock inputs, most of which have a market price. Fixed costs relate to flow inputs, often provided from household resources and hence free of charge. Their opportunity costs are not easily assessed.

Variable costs can usually be allocated fairly easily, to individual enterprises, except where there is mixed cropping. Just as there are various alternative ways of allocating inputs to components of mixed crops, none of them wholly satisfactory, so, too, is there a choice of methods of allocating variable costs. For any given enterprise the gross margin is the difference between gross output and variable cost.

Enterprise gross margin
= enterprise gross output
− enterprise variable costs

Once again crop gross margins are usually presented on a per hectare (or per tree) basis while livestock gross margins are presented per head or per livestock unit. The total sum of all the enterprise gross margins gives the total farm gross margin.

If fixed costs do not alter much with changes in production, then where total gross margin can be increased, farm profit or surplus will rise. If the increase in gross margin can be achieved with the existing supply of fixed resources and hence the existing level of fixed costs, profit will be raised by exactly the same amount as the gross margin. For this reason it is possible to plan changes in the farm system in terms of gross margins alone and leave fixed costs out of the calculation. In fact, in many parts of Africa, the family farmer does not incur explicit fixed costs. He pays no rent, nor wages to his family who make up his regular labour force, he has hardly any buildings and equipment and does not borrow much capital. Practically all the African farmers' costs are variable. This means that practically the whole of the total gross margin represents family or social income.

Thus one useful method of completing the farm business analysis is to compute the enterprise gross margins per unit of limiting resource. In some cases, where land of a certain type (eg irrigated land) is limited, comparisons of gross margin per hectare are useful. In other cases, comparisons of gross margins per man–day of peak labour may be more appropriate.

Alternatively, given that there is some expenditure on fixed costs, of wages for instance, land rents or machinery operating costs, these together with an estimate of the depreciation of machinery and equipment, may be subtracted from the total farm gross margin to estimate net farm income; thus

Net farm income = total gross margin
− explicit fixed costs

If income from off-farm activities, including remittances and wages earned from off-farm employment, can be estimated, it is useful to add these to the net farm income to give an estimate of *total household income*. This estimate of the household income from all sources may be divided by the household size (measured in standard consumption units) to arrive at the income per consumption unit.

Other analyses

(i) *Financial analysis*

It may be useful to carry out a separate analysis to investigate the financial position of the farm household. Such an analysis is concerned solely with cash receipts and expenditure. The total of farm receipts from crop and livestock sales, minus total expenditure on the purchase of farm inputs gives the farm cash surplus.

Farm cash surplus = Total farm receipts
− total farm expenditure

Where credit is used, the results may be further adjusted to allow for loans received and debts repaid, thus

Farm cash surplus after financing
= Farm cash surplus
+ farm loans received
− repayment of principal and interest

Finally cash income from off-farm activities may be added to give the *household net cash income*. This is a measure of the amount of cash available for meeting all payments not relating to the farm. It is a less comprehensive measure of welfare than the total household income. None the less, it may be useful to consider the financial position separately from income in kind which is consumed within the household.

(ii) *Cash flow analysis*

For long-term investments such as permanent crops, there are obvious problems in obtaining the long series of costs and returns data needed for a comprehensive evaluation. The only possible sources of records, over the lifetime of cocoa or oil-palms, are research station reports or long-established plantations. However, information may be obtained from a farm survey on the annual costs and benefits at different stages of the life cycle, from different plots. Thus, it may be possible to build up or synthesize a lifetime profile of costs and returns by combining data from different aged plots.

For purposes of evaluation it is necessary to calculate the annual cash flow, meaning the difference between total revenue and total cost for the enterprise in each year. Cash flows differ from gross margins in that no attempt is made to estimate annual depreciation or appreciation of assets. Instead, the full cost of any capital investment is recorded in the year when it occurs. Similarly, if assets are sold, the sale price is recorded in the year of sale. Costs of labour, even family labour, must be estimated and subtracted in estimating cash flows. Discussion of methods of evaluating the resultant stream of cash flows is deferred until Chapter 15 on Investment Appraisal.

(iii) *Livestock productivity*

In addition to the estimates of gross margins per head or per livestock unit, already discussed, further livestock productivity measures are desirable. More specifically it is useful for problem diagnosis, and herd or flock growth modelling purposes to estimate
(a) reproduction rates, which may, in turn, depend upon age at first parturition, parturition interval or parturition rate and average litter size,
(b) age-specific mortality rates, and
(c) daily liveweight gain.

Calculation of these measures is, of course, only possible if the necessary data have been recorded. The crude reproduction rate may be estimated as the total number of live births during the year divided by the average number of breeding females in the herd or flock. If the number of parturitions (P) is recorded then the parturition rate (R) is given by dividing by the average number of breeding females (N).

$$R = P/N$$

The mean parturition interval (I) in days is obtained as

$$I = 365/R$$

While average litter size (L) is the number of live births per parturition

$$L = B/P$$

where B = total number of live births.

The crude mortality rate is simply the ratio of the number of mortalities to the mean number of animals. Age-specific mortality rates are calculated in the same way for specific age cohorts.

Two general points may be noted. First, given that numbers of animals are constantly changing over time, frequent recording is needed to arrive at accurate estimates of average numbers used in estimating reproduction and mortality rates. The second point is that, since individual household flocks and herds are relatively small, there may be many gaps in the estimates of age-specific mortalities and the variation between households in all these measures is likely to be large.

Comparative analysis

The need for comparing adopters with a control group of non-adopters has already been emphasized. However, much may be learned in the diagnosis of constraints and identification of improvements by comparing performance on different farms, given that some farmers are more innovative and successful than their neighbours. Comparisons of the farming systems of the more successful with those of the less successful may help to identify where the critical difference lies. For an example of detailed analysis of this kind see Upton & Petu (1966), and Upton (1964).

Differences may lie in the inherent abilities of the household decision-makers or in their resource endowments; in which case the less successful family may be unable to emulate their more successful neighbours. Nevertheless, it is useful for the researcher to discover whether this is the case. Indeed, it may be possible to promote institutional change which will improve the resource base of poorer households. In other cases, useful indigenous innovations may be identified as a result of comparative analysis.

Further reading

Bayliss-Smith, T. P. (1982). *The Ecology of Agricultural Systems*, Cambridge University Press

Clark, C. & Haswell, M. (1964). *The Economics of Subsistence Agriculture*, London, Macmillan

Dillon, J. L. & Hardaker, J. B. (1980). *Farm Management Research for Small Farmer Development*, Rome, FAO, Agricultural Services Bulletin 41

FAO (Food and Agriculture Organisation of the United Nations) (1983). *The FAO Farm Analysis Package: A General Introduction*, Rome, FAO

Freund, J. E. (1979). *Modern Elementary Statistics*, 5th edn, Englewood Cliffs, New Jersey, Prentice-Hall

Nie, N. H., Hull, C. H., Jenkins, J. G., Steinbrenner, K. & Bent, D. H. (1975). *Statistical Package for the Social Sciences*, 2nd edn, McGraw-Hill

Upton, M. (1964). A development of gross margin analysis, *Journal of Agricultural Economics,* **16**(1) 111

Upton, M. & Petu, D. A. (1966). A study of farming in two villages in the middle belt of Nigeria, *Tropical Agriculture*, Trinidad, **43**(3) 179

13

Production functions

Data for production function analysis

The production standards discussed in the last chapter are useful for comparing *average* products of specific resources on different farms or in different enterprises. However, in theory we would expect *marginal* products to be more important in determining the economic optimum level of production, combination of resources and of enterprises. If we are to use the marginal approach to decision-making we must first establish the production function relating output to different levels of inputs. Hence we need a series of observations at different levels of input.

Several observations are needed even in the simplest case where there is one single variable input and one single product and the relationship is assumed to be linear (a straight line). This is illustrated graphically in Figure 13.1 showing hypothetical data relating nitrogen fertilizer input to maize yield. In Figure 13.1(a) we have only one single observation of input and output; there is only a single point on the graph. Obviously, any number of straight lines, all with different slopes, could be made to pass through this single point. The marginal product cannot be estimated. It should be noted, however, that the average product is easily obtained by dividing output by input.

In Figure 13.1(b), where there are two observations and hence two points, there is only one straight line which will pass through both. The slope and hence the marginal product per unit of nitrogen fertilizer on maize can be estimated. However, with only two points, we have no way of assessing the reliabililty of our estimate of the slope. If there are errors of measurement of either inputs or outputs or if there are other factors influencing output which we

have not taken into account, any further observations which are made may not lie close to the line at all. In order to make a more reliable estimate and to assess its reliability, many observations are needed, as in Figure 13.1(c). Generally speaking, the more points that are available, the greater the reliance that can be placed on the estimated relationship. In practice, the problem is further complicated where there are several variable inputs and curved relationships. More observations are needed where there are several inputs which can be varied and where various different curved relationships are considered possible.

A suitable series of observations may be obtained from controlled experiments if they were designed with this object in mind. To fit a function to experimental data, several levels of each input treatment must be included in the experiment. This is often not the case, many experiments having been designed to test whether a particular treatment, sometimes at a single level, has a 'significant effect'. However, more and more researchers in crop and livestock production are realizing the benefits of designing their experiments to measure the slope of the production response curve.

A production function may be fitted to survey data, the results for each farm representing a single observation. Various problems arise with this approach, since none of the variables are controlled as they are in an experiment. In particular, environmental conditions and managerial ability vary from one farm to another. Furthermore, since practically all inputs may vary from farm to farm some aggregation of both inputs and outputs may be needed.

A production function can only be fitted to data for a single farm, such as a unit farm, if results for several years' operation at different input levels are

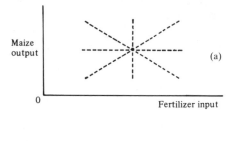

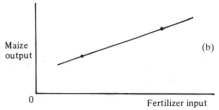

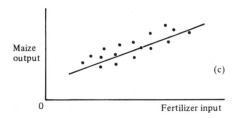

Fig. 13.1 Fitting a straight line, (a) Single observation, (b) Two observations, (c) Many observations

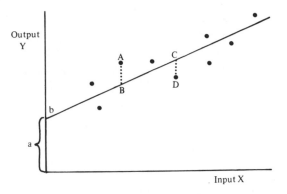

Fig. 13.2 The straight line and residual errors

available. These would then represent the series of observations, in this case a time-series. Such a set of time-series data is unlikely to extend over many years so the scope for production function analysis of single farm data is limited.

Methods of estimating the slope

The production function may be described, as it has been in earlier chapters, by a set of tabulations of specific inputs and the related outputs. However, this is a rather clumsy method and no information is given regarding intermediate points. The economic optimum cannot be estimated precisely, but only to the nearest unit of input. It is therefore customary to attempt to relate inputs and outputs by means of a smooth curve.

Where only one input and one output are involved, the observations can be plotted on a graph as in

Figure 13.1(c) and a curve fitted to these points by freehand drawing. Some subjective judgement is involved in drawing the shape and slope of the curve, but the accuracy must depend upon how widely the points are scattered. Alternatively any one of a variety of more systematic mathematical methods can be used.

The most widely used and best known of these mathematical techniques is the method of least squares regression. In its simplest form, it is used to fit a straight line, such as $Y = a + bX$ in Figure 13.2, where a and b are the unknown values which are to be estimated and X is the level of input and Y the level of output. The object is to make the sum of the squared vertical deviations between the recorded points and the line as small as possible. These deviations such as AB or CD in Figure 13.2 are also known as residual errors, or residuals. It will be noted that a is the value of Y when the level of input X is zero, and that b is the slope of the line, and hence represents the marginal product of X.

The idea is inherently attractive. We wish to minimize the residual errors, or the variation about the line, because this implies that the line is a good fit to our points. If the residual variation were large, the line would not represent the information in a very satisfactory way. One way of measuring the variation is in terms of the mean absolute deviation from the line (see Chapter 5 for discussion of a comparable measure of the variation about the mean). However, the sum of the squares of residuals is a more useful measure, while minimizing the sum of squares generally gives a better fit to all the data points than does minimizing the sum of absolute deviations.

It is possible to compare different lines, which have been fitted freehand, by comparing the sums of squares of residuals, *but* the mathematical technique of least squares regression enables us to estimate the specific values of a and b which minimize the sum of squared residuals. The formulae are

$$b = \sum x_i y_i / \sum x_i^2$$

and

$$a = \overline{Y} - b\overline{X}$$

where $\overline{X}$ and $\overline{Y}$ represent the means of the two variables and x_i and y_i represent the deviations from these means. Thus once we have decided on the general shape of the relationship between X and Y, least squares regression enables us to find the line of best fit objectively without having to rely on our personal judgement.

The output Y is referred to as the dependent variable, while the input X is called the independent variable. Multiple regression is the extension of this analysis to include cases where there is more than one independent variable. Thus more than one variable input may be taken into account.

The simplest case of multiple regression, is that of multiple linear regression with two independent variables. In that case the equation is of the form

$$Y = a + b_1 X_1 + b_2 X_2$$

which represents a plane in three dimensions. The coefficient b_1 is the slope in the X_1 direction, or in other words the effect of varying X_1 when the other independent variable X_2 is held constant. Similarly, b_2 measures the effect of varying X_2 when X_1 is held constant. Estimation of the values of b_1 and b_2 requires solution of a set of simultaneous equations (known as the Normal Equations).

$$\sum x_{1i} y_i = b_1 \sum x_{1i}^2 + b_2 \sum x_{1i} x_{2i}$$

$$\sum x_{2i} y_i = b_1 \sum x_{1i} x_{2i} + b_2 \sum x_{2i}^2$$

$$Y = a + b_1 \overline{X}_1 + b_2 \overline{X}_2$$

(eg see Wonnacott & Wonnacott, 1970).

Where there are more than two independent variables, the relationship cannot be imagined in three dimensions. Although the methods of estimation are based on the same principles, they are of course, more complicated and use of a computer is recommended.

Regression analysis

The method of least squares enables us, not only to find the line of best fit, but also to measure how good a fit it is. Let us first note that the variation about the mean of Y, can be measured by the variance, which we may write as

$$\sum y_i^2 / (n - 1)$$

But each deviation y_i consists of two parts, the deviation of the *predicted* value of Y_i from the mean written as $\hat{y}_i$, and the residual e_i (see Figure 13.3). If, as is normally assumed, there is no correlation between y and e, then

$$\sum y_i^2 = \sum \hat{y}_i^2 + \sum e_i^2$$

(the variation in Y) = (the variation 'explained' by the regression) + (the residual variation).

These values may be calculated and used to derive the coefficient of determination (R^2)

$$R^2 = \frac{\sum \hat{y}_i^2}{\sum y_i^2} = \frac{\text{proportion of the total variation}}{\text{explained by the regression.}}$$

This clearly is a measure of goodness of fit. A value of 1 for R^2 means that all the variation is explained, or

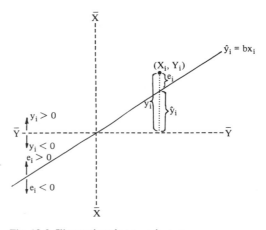

Fig. 13.3 Illustration that $y_i = \hat{y}_i + e_i$

that the regression predicts the Y values exactly. A value of zero means there is no association.

The analysis can be taken further by assuming that each predicted value of Y and indeed the value of 'a' and each of the 'b' coefficients, is an 'expected value'. Thus a standard error can be estimated in each case and used to construct a confidence interval or to test hypotheses. It is usual practice, for instance, to test the hypothesis that the true value of each coefficient is zero; there is *no* association. If the hypothesis can be rejected we say the coefficient is 'significant'.

A word of warning is needed, however. Strictly speaking the estimation of confidence intervals and hypothesis testing is only valid when appropriate random sampling methods have been used. Furthermore, if the sample is small, say less than 30 cases, we have to assume that the sample is drawn from a normally distributed population.

Forms of function

I. *The linear function* So far we have been discussing the linear function, which is based on the assumption that the inputs and outputs are all related by straight lines. This means that the slope of each relationship is constant and hence that the marginal product is constant. It makes no allowance for diminishing

marginal returns so there can be no economic optimum. The total and marginal product graphs for this function are shown in Figure 13.4. It is assumed that the relationships between inputs are constant (see isoquants illustrated in Figure 13.5), which means that all inputs are perfect substitutes for each other with constant rates of technical substitution. This is obviously nonsense since it would mean that the least-cost combination of resources would consist of one *single* resource input, namely the cheapest per unit of output.

Clearly, the linear function is not satisfactory on theoretical grounds and its use can only be justified on the basis of the ease of fitting it by least squares regression and as an approximation over a limited range of values.

Fortunately, our standard regression procedure can be applied to curved relationships, provided that the variables can be redefined, or the equation transformed to make it linear. The two functional forms most commonly used in this way are the quadratic, or second degree polynomial, and the Cobb–Douglas or power function.

II. *The quadratic function* The equation for the quadratic function, in the case of two variable inputs, is:

$$Y = a + b_1 X_1 + b_2 X_2 - b_3 X_1^2 - b_4 X_2^2 + b_5 X_1 X_2 \tag{1}$$

where Y is the level of output and X_1 and X_2 the level of each of the two variable inputs. Additional variables are formed by squaring the values of X_1 and X_2 and by forming the product of these two. Thus X_1^2

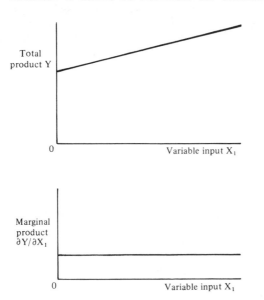

Fig. 13.4 The linear function for Y on X_1 with X_2 fixed

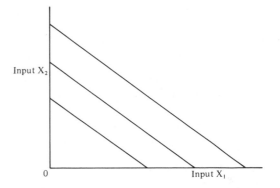

Fig. 13.5 Isoquants for the linear function

could be thought of as a new variable X_3, likewise $X_2{}^2$ could be thought of as X_4 and X_1X_2 as X_5. For instance, if for a particular observation in a fertilizer trial X_1 is 25 kg of ammonium sulphate and X_2 is 50 kg of superphosphate, then X_3 is $25^2 = 625$, X_4 is $50^2 = 2500$ and X_5 is $25 \times 50 = 1250$. The values for these new variables are calculated for other observations in the same way. The unknown values of a and all five b's can then be estimated by multiple regression of Y on the five X variables.

In the quadratic equation (1) above, b_1 and b_2 measure the slope of the curve at zero input. They are normally positive, showing a positive production response to increasing inputs of variable factor from zero upwards. On the other hand b_3 and b_4 measure the rate of change in the slope of the response curve. Thus if there are diminishing marginal returns b_3 and b_4 should have negative signs as shown in equation (1). The interaction between the two variable inputs occurs in the last term of the equation. It is usually positive, meaning that the two inputs are more productive when used in combination, but negative or zero interaction may exist where diminishing marginal returns hold true for both factors. The constant

a is the output obtained when X_1 and X_2 are both zero. It therefore represents the output from the mix of fixed resources and may sometimes be zero.

The marginal products are obtained by differentiation

$$\text{so } MP_1 = \partial Y/\partial X_1 = b_1 - 2b_3X_1 + b_5X_2$$
$$MP_2 = \partial Y/\partial X_2 = b_2 - 2b_4X_2 + b_5X_1$$

The typical shapes of the total and marginal product graphs for this function are shown in Figure 13.6. This function can show diminishing marginal returns and even negative ones. There is then a technical optimum beyond which the total product falls. An economic optimum occurs where the marginal value product equals the unit factor cost. A quadratic function can show increasing marginal returns if b_4 and b_5 are positive, but it can never show both increasing marginal products at low levels of input and decreasing marginal products at higher levels of input in the same equation. Furthermore, at very high levels of input and possibly for very low ones too, this function may predict negative total products which are clearly impossible. In such a case the function ceases to be meaningful for very high or very low levels of input.

The isoquants and two possible expansion paths are shown in Figure 13.7. The isoquants are generally convex to the origin at low levels of input so they show diminishing rates of technical substitution. Thus the least-cost method of production is likely to include both variable inputs.

The isoquants may cut the axes of the graph, as is true for the lower output curve in Figure 13.7, which

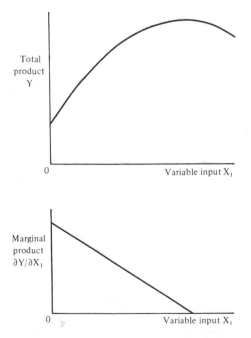

Fig. 13.6 The quadratic function for Y on X_1 with X_2 fixed

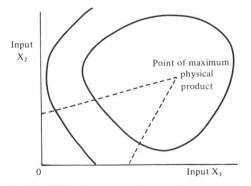

Fig. 13.7 Isoquants for the quadratic function

cuts the X_1 axis. It is implied that this level of output can be achieved by using the first variable input alone and none of the second. The expansion paths are straight lines which do not necessarily pass through zero, but converge to the point of maximum physical product. This means that the least-cost combination of resources, that is the optimum ratio of X_1 to X_2, varies according to the level of output.

A possible disadvantage with this form of function is the large number of b values, or regression coefficients, which must be estimated for a given number of variable inputs. Thus, with a linear function the number of regression coefficients is the same as the number of variable inputs, but for a quadratic function, one variable input involves two regression coefficients, one for the X term and one for the X^2 term. Furthermore, the number of coefficients increases more rapidly than the number of variable inputs. In equation (1) with two variable inputs there are five coefficients. With three variable inputs there are ten coefficients if all possible interaction effects are estimated. Hence, if many inputs are allowed to vary in the production function the quadratic function may become very large and cumbersome and may require a very large number of observations for reliable estimation.

In summary, the advantages of this form of function are that it is relatively easy to estimate and that it may show diminishing marginal returns. The possible disadvantages are that it cannot show both increasing and diminishing marginal returns in a single response curve, that it may give negative total products for very high or very low levels of input and that it becomes complex if many variable inputs are included. It is most commonly used for analysing experimental results where there are few variable inputs but where zero variable inputs do not necessarily yield zero output.

III. *The Cobb–Douglas function* The equation for this function, in the case of two variable inputs, is:

$$Y = AX_1^{b_1}X_2^{b_2}. \qquad (2)$$

It is named after two men called Cobb and Douglas who together used it for a production function study in America in 1928. The a and b coefficients are estimated by converting all the variables measured,

both inputs and outputs, into their logarithms and then using ordinary linear least squares multiple regression on these logarithms, thus:

$$\log Y = a + b_1 \log X_1 + b_2 \log X_2. \qquad (3)$$

Equation (2) is simply the antilog of equation (3) so that A is a multiplicative constant and the antilog of a. The Cobb–Douglas function is sometimes known as a logarithmic function or even more precisely a double-log function, to distinguish it from other functions where only one side of the equation is transformed into logarithms. The equation is easily extended to include more variable inputs.

The marginal products are given by

$$MP_1 = \partial Y/\partial X_1 = b_1 AX_1^{(b_1-1)}X_2^{b_2} = b_1 Y/X_1$$
$$MP_2 = \partial Y/\partial X_2 = b_2 AX_2$$

That is (marginal product) = b × (average product). The average product varies however, depending on the level of input, so it is usually estimated at the average level. Where there are diminishing marginal returns, b_1 and b_2 are less than 1. A b coefficient of exactly 1 implies constant marginal returns and one greater than 1 implies increasing returns. Since the effect of scale is measured by the sum of elasticities of response for all inputs, the Cobb–Douglas function may be used to estimate returns to scale, *provided that all inputs have been included in the function*. The sum of the b coefficients then gives an estimate of returns to scale. If the sum is greater than 1 than there are increasing returns and if the sum is less than 1 there are decreasing returns.

The typical shapes of the total and marginal product curves for the Cobb–Douglas function are shown in Figure 13.8. Provided that the b coefficient is less than 1, the response curve shows diminishing marginal returns. However, negative marginal returns are not possible so there is no technical optimum or maximum total product. In fact the total and marginal product curves tend to flatten out into an almost straight line at high levels of input. As a result of this, if the economic optimum occurs at a fairly high level of input, the Cobb–Douglas function may give an overestimate of the economic optimum. It will be noted that at zero level of input the output is also zero. This is invariably the case with a Cobb–Douglas function, unlike the quadratic. This is because the

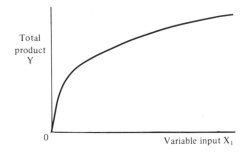

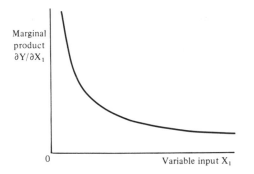

Fig. 13.8 The Cobb–Douglas function for Y on X_1 with X_2 fixed

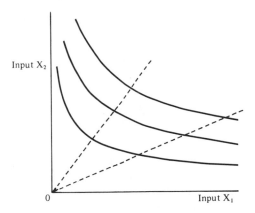

Fig. 13.9 Isoquants for the Cobb–Douglas function

variable inputs in equation (2) are multiplied together, so if any one of them is zero, the product must also be zero.

The isoquants shown in Figure 13.9 are again convex to the origin, showing diminishing rates of technical substitution. They never cut the axes however, thus implying some complementarity between resources. It is impossible, according to this function, to produce any product without some of each resource; one can never substitute entirely for another. The expansion paths are straight lines passing through zero. This implies that the least-cost combination (ratio) of resources is the same at all levels of output. Once the optimum combination has been found this can be increased to scale.

The advantages of the Cobb–Douglas function are that it is easy to estimate, it may show diminishing marginal returns and can also be used to estimate returns to scale. Possible disadvantages are that it cannot show both increasing and diminishing marginal returns in a single response curve, that it does not give a technical optimum and may lead to over-

estimates of the economic optimum. The implication of zero output at zero input may be unacceptable in some instances. For example, some crop product is usually obtained even when no fertilizer is applied. The implication of a constant elasticity of response at all levels of input may also restrict the usefulness of this function. It is commonly used for analysing survey data, where many variable inputs are included and it is hoped to measure returns to scale.

Production functions in practice

The mathematical functions which have been discussed are, of course, not the only ones which could have been used to describe production relationships (see Dillon, 1977). Nevertheless, they are the ones most commonly met with in practice. The choice of function involves a certain amount of subjective judgement as does the choice of variable inputs to include in the function. It is possible to compare the values of R^2 obtained or use statistical tests of whether one model is an improvement on another, but since there is always some residual variation in output we cannot prove conclusively that one particular function is the correct one. Economic data do not obey exact laws; hence it is impossible to predict output exactly for a given combination of inputs. However, as we have seen, the quadratic and the Cobb–Douglas functions have certain desirable characteristics from a theoretical point of view.

Nevertheless the frequency with which these two

forms are used is probably due, in part, to the ease with which they can be estimated.

Various problems arise in applying production function analysis to any data. One general problem is that in theory the production function represents an instantaneous relationship between inputs and output. In practice there is invariably a delay between the use of the input and the output response. Thus, in practice, the relationship must be measured over a period of time, usually an agricultural year. However, this is not satisfactory where the delay is longer than a year, as is true for many capital investments. Furthermore, capital inputs are usually chunky or indivisible so a smooth mathematical function cannot be used to describe response. The introduction of capital inputs into a production function therefore raises considerable problems which have not been completely resolved (see Upton, 1979).

An associated problem is that of allowing for risk and uncertainty. The economic optimum represents the level of output which will yield the highest profit *on average*, but there must be considerable uncertainty about the outcome of any productive activity in a particular season on a particular farm. In fact, as we have seen, most farmers make some efforts to avoid risks and are willing to give up some profit for this purpose. Thus the most attractive level of output for most farmers is likely to be somewhere below the economic optimum, since they will discount the potential marginal returns for risk. Furthermore, since farmers vary in their aversion to risk, then the 'most attractive level of output' will vary between farmers. This problem also has not been completely resolved. Other problems are associated with particular studies, whether they are 'experimental' or 'survey', so one example of each type of production function study will be discussed in more detail.

Technical experimentation: response of sorghum to fertilizers in Northern Nigeria

This study was based on trials to investigate the effect of nitrogen and phosphate fertilizer on the grain yield of sorghum. (See Goldsworthy, 1967.) A total of 154 trials were carried out throughout Northern Nigeria during the five-year period 1960–64. Nine fertilizer combinations were tested in the trials, namely three

levels of ammonium sulphate (nitrogen fertilizer) and three levels of superphosphate, using $0:56:112$ lb of each per acre (approximately $0:63:125$ kg per hectare), in all possible combinations. At each site the nine treatments were repeated four or five times, so there were four or five observations for each fertilizer combination at every site.

It would have been possible to fit a production function to the observations for a single site but in this case the research worker made use of all 154 trials. He ended up with five response areas. Three of the five areas, namely those in the southern part of Northern Nigeria, were grouped together because the responses in each of them were too low to be of economic interest. The production functions estimated for the other two areas were:

Area I (covering most of the northern part of the country)

$$Y = 930.58 + 77.17X_1 + 92.17X_2 + 2.17X_1^2 - 51.83X_2^2 + 9.25X_1X_2$$

Area II (roughly associated with loess soils of the Zaria plain)

$$Y = 1100.65 + 44.67X_1 + 223.67X_2 - 3.00X_1^2 - 124.00X_2^2 + 45.50X_1X_2$$

where Y is yield in lb of sorghum grain per acre,
X_1 is level of nitrogen application, (1, 2 or 3)
X_2 is level of phosphate application, as defined above.
(Coefficients for X_1, X_2 and X_2^2 significantly different from zero in each case.)

These equations are typical quadratic functions except that for Area I, the value of X_1^2 is multiplied by a positive coefficient of $+2.17$. This implies increasing marginal returns to nitrogen.

The economic optimum level of fertilizer use was not estimated since it appeared that for these two areas it would lie well above the maximum level of application used in the trials. The author rightly considered that it would be dangerous to estimate responses at levels of fertilizer use which had not been studied. However, the equations were used to draw isoquants and estimate the least-cost combinations of nitrogen and phosphate fertilizers at different levels of output. Generally, the expansion path was found to lie nearly parallel to the nitrogen axis, showing that increasing product per acre is most cheaply achieved by increasing the proportion of nitrogen in the fertilizer combination.

This example illustrates several of the typical advantages and disadvantages of the use of experimental data for production function analysis. First, there is the advantage that the inputs can be set at controlled levels so that observations may be well spaced out over the response curve and all combinations of the inputs can be investigated. This is not usually the case with survey data. Secondly, as many observations may be obtained as desired, and with a sufficiently wide range of observations it may be possible to test a number of different functions to see which of them gives the best fit. A third advantage is that most of the variables not included in the function may be held constant, for all the treatments. Thus the effects of differences in soil type can be virtually eliminated and the standard of management, the labour inputs and the timing of all operations are identical for all the different levels and combinations of fertilizer use.

The disadvantages of the experimental approach are associated with the fact that relatively few inputs can be varied satisfactorily in any one experiment. The majority of inputs are held constant at a specific level. This means that the interactions between the fixed inputs and the variable inputs cannot be measured, and these interactions may be very important. For instance, in the work described, if the standard of management, the labour inputs or the timing of operations had been different the responses of sorghum yield to nitrogen and phosphate fertilizers might also have been different. Unfortunately, the standard of management of experimental trials is usually much higher than that found in practice on farms.

For this reason, in general, response measured under experimental conditions significantly exceeds the response achieved under farm conditions. The experimental results are therefore of limited value in advising farmers or in planning their farms.

An additional disadvantage of the experimental approach is that the true opportunity cost of inputs may not be known. This is because each experiment is usually concerned with only one single enterprise. Information may not be available on alternative activities. Thus, in the example quoted, the fertilizers and the sorghum yield were all valued at current market prices. Now the use of fertilizer represents a short-term capital investment and it is possible that other forms of investment would yield a higher return than the market price of the fertilizer. This means that the opportunity cost of the fertilizer would be greater than its market price. Hence this type of analysis could give misleading results. In terms of market prices fertilizers might appear profitable when, in terms of opportunity cost, they might not.

Farm survey analysis: smallholder farming in Zimbabwe

This analysis is based on a survey of small farms on what was the Chiweshe Reserve in Zimbabwe. Although production function analyses of farm surveys have been made in other parts of Africa this one has been described and discussed in particular detail (see Johnson, 1969 & Massell, 1967).

The sample survey was originally conducted during the 1960–61 cropping season and included 118 farms in all. It is claimed that 'in terms of crop production, the 1960–61 season was approximately average for Chiweshe [and] the area sampled appears to be reasonably representative of the reserve as a whole.' A large variety of crops are grown and some livestock are kept but the analysis was restricted to the three main crops of maize (corn), millet and groundnuts (peanuts). For the production function analysis all farms that had incomplete crop production data or that did not grow all three crops were deleted, leaving a final sample of fifty-six farms in the analysis.

A Cobb–Douglas function was used to relate the output of each crop to the set of observed inputs used in producing the crop. The function can be written, using our notation,

$$\log Y = a + b_1 \log X_1 + b_2 \log X_2 \\ + b_3 \log X_3 + b_4 \log X_4 + b_5 \log X_5 \\ + b_6 X_6 + b_7 X_7 + b_8 X_8$$

where Y = physical output of the particular crop (maize, millet or groundnuts),
X_1 = area of land used for that crop,
X_2 = man–hours of labour used in weeding that crop,

X_3 = weight of chemical fertilizer applied to that crop (plus a constant),

X_4 = weight of organic manure applied to that crop (plus a constant),

X_5 = value of farm implements owned, at undepreciated initial cost,

X_6 = soil type, red loam or sandy soil,

X_7 = skilled farmer, yes or no,

X_8 = semi-skilled farmer, yes or no.

An extra term was added to represent the residual error but since we assume that on average the residual error is zero we can omit it from the equation.

This equation differs from the simple Cobb–Douglas function described earlier in that more variable inputs are included. Furthermore, some of the X variables do not represent quantities of a particular input, but simply take one of two values. For instance X_6 representing soil type takes the value of 1 for red loam or 0 for sandy soil. Variable X_7 takes the value of 1 for a skilled farmer or 0 for a semi-skilled or unskilled farmer. Likewise variable X_8 takes the value of 1 for a semi-skilled farmer or 0 for any other. Such variables are known as 'dummy variables' and are used to estimate the effects of factors which are not easily measured as physical quantities. These dummy variables are not converted into logarithms.

Before discussing the results, the variables included in the function will be considered in a little more detail.

Output is measured in physical units of weight harvested. To compare marginal value products, however, physical outputs must be multiplied by a measure of value.

Average market price paid in the area is used. For maize and groundnuts this was the official Grain Marketing Board price but for millet it was the local market price.

Land is measured in terms of the area devoted to each crop, but land is not assumed to be all of the same quality. The effect of soil type is, at least in part, allowed for by means of the dummy variable X_6.

Labour was provided by members of the farm family. For each crop, labour inputs were recorded for each of the major operations: applying manure, planting, weeding and harvesting. Because labour appeared to be a limiting factor only at weeding time, the number of weeding-hours is used as the labour

variable. Hours worked by children are weighted by one-half.

Two kinds of fertilizers were used, chemical and organic, but only on the maize land. The variables X_3 and X_4 measure the quantity of fertilizer or manure input plus a constant, which in both cases is 100. A constant must be added before converting to logs since some farmers did not use any on their maize and yet still obtained some output. (Note that log 0 does not exist.) This implies that fertilizers are not essential inputs as there is some natural fertility in the soil. The constant may be assumed to represent the natural fertility but the choice of its value is quite arbitrary.

Fixed capital consisted of relatively simple farm implements such as an ox-drawn plough or cultivator. As an index of a farmer's fixed capital inputs, the value of farm implements at undepreciated initial cost is used. This index omits the services of draught animals and investment in the land, neither of which was recorded in the survey. Furthermore, no account is taken of the current condition of the implements.

Managerial inputs are included in the function by means of the dummy variables X_7 and X_8 which are based on a rating of farmers by the government agricultural extension service. Thus farmers who receive advice from the extension service are classified into three categories: co-operators, plot holders and master farmers. A co-operator is any farmer who uses fertilizer, carries out some crop rotation and plants his crops in rows. A plot holder is a farmer who is under tuition by an extension worker to become a master farmer. A master farmer is one who has gone through the plot holder stage and has reached specified higher standards of crop and animal husbandry as laid down by the agricultural department. Of the fifty-six farms in the final sample there were three master farmers, four plot holders and fourteen co-operators. Owing to the small numbers, master farmers and plot holders are combined into a single group of 'skilled' farmers. The co-operators are referred to as 'semi-skilled' and the remaining thirty-five farmers as 'unskilled'. The coefficients b_7 and b_8 are a measure of the contribution to output of 'skill' and 'semi-skill' relative to lack of skill.

The results of the analysis show that there is considerable variation in output which is not explained by the production function. For groundnuts and millet less than half the total variation in output is explained by the analysis, so the authors rightly warn that the results must be interpreted with caution ($R^2 < 0.5$). The only coefficients significantly different from zero were:

for maize – land, soil type, chemical fertilizer and organic manure,

for groundnuts – fixed capital and skilled management,

for millet – land and weeding labour.

The sum of the elasticities of response is nearly 1 for maize and millet, thus suggesting constant returns to scale for these two enterprises. However, the sum of elasticities is only 0.753 for groundnuts, which implies decreasing returns to scale. This may be due to the omission of some important factor, such as labour quality, that should enter the groundnut production function.

The estimated marginal value products of the variable inputs from each of the three crops are given in Table 13.1, converted from the original units into £ sterling and metric system physical measures.

These results suggest which variable inputs are most profitable to expand. There is no possibility to bring more land under cultivation as farmers use all the arable land. Land appears to be a limiting factor. The marginal value product of labour is low in relation to the hourly wage rate in paid employment. Because of the low return to labour on the farm, many farmers spend a considerable part of the year away from the reserve working for wages. The return to chemical fertilizer does not appear adequate to justify much increase in its use. On the other hand it is suggested that the unit factor cost of organic manure is very low, virtually only the labour cost, so the marginal product is a return to labour. As an average of 16 hours was spent applying a ton of organic manure the return to this labour is about 8p per hour as against only just over 1p per hour for weeding. However, livestock numbers may be an effective constraint on the amount of manure available. The return to capital is low and the results suggest that the area is overcapitalized with respect to implements.

With regard to soil type, the benefit from farming on the red loam soils rather than sandy soils appears

Table 13.1. *Estimated marginal value products*

Input	£ per unit of measure		
	Maize	Groundnuts	Millet
Land per hectare	3.13	3.04	4.40
Labour per weeding hour	0.005	0.012	0.015
Chemical fertilizer per £ cost	1.69	—	—
Organic manure per metric ton	1.31	—	—
Fixed capital per £ cost	—	0.087	0.025
Soil type per hectare (advantage of red loam over sandy soil)	0.88	0.21	1.07
Skilled farmer	1.35	1.52	−1.05
Semi-skilled farmer	−0.35	0.81	0.29

From Massell (1967). *Farm Management in Peasant Agriculture: An Empirical Study.* They differ slightly from those given in some of the other reports of this research.

greatest for millet and lowest for groundnuts. The estimates for managerial skill measure technical efficiency only and do not reflect differences between farmers in allocative efficiency. For instance, the figure for skilled farmers in the maize production function is £1.35; this means that on average for a given level of all other inputs, skilled farmers obtained £1.35 more maize output than unskilled farmers. There are some unexpected results in that skilled farmers obtained *lower* returns from millet production than unskilled farmers, and the same is true for semi-skilled farmers in maize production. It is suggested that this may reflect possible shortcomings in the government rating scheme which tended to focus on maize and groundnuts, or it may be the result of small sample size. The study showed that the quantities of all other resources used are related to managerial skill. Skilled farmers use more land, labour, fertilizer, manure and fixed capital than the semi-skilled, who in turn use more of all these resources than the unskilled.

The analysis is also useful in suggesting how profits may be increased by reallocation of resources. The marginal value productivities of both land and labour are highest in growing millet, suggesting that profits would be raised by shifting resources from maize and groundnuts into millet production. However, the resulting gain is estimated to be relatively small. In so far as the farmers may have objectives other than profit maximizing, such as self-sufficiency, the existing allocation of resources may be satisfactory. This, of course, is only true *on average*. Some individual farmers might benefit considerably from reallocation of resources.

The problems of applying production function analysis to farm survey data arise from the fact that it is impossible for the researcher to control any of the variable inputs. At least, he may restrict his study to farms of a certain size range or type, but he cannot set any inputs at selected fixed levels, or arrange that the inputs are varied independently of each other. This still need not create serious problems if it could be assumed that the quantities of inputs used on different farms varied at random. Unfortunately this is not the case, since the quantities of resources used are largely the result of conscious human decisions. For instance, if the sample of farmers all operate on the same production function, all pay the same prices for inputs and all operate at the economic optimum, then they would all use exactly the same quantity of each input and produce exactly the same output. Although there would be a large number of farms in the sample, they would all be operating at the same point on the production function. It would be impossible to draw or estimate the form of the function as in Figure 13.1a.

In fact the problem need not arise in this extreme form since some inputs are fixed at different levels on different farms, such as the supply of land, the family size, or the managerial ability of the farmer. Alternatively, there may be variation between farms in the prices paid for resources. However, it still remains true that the inputs do not vary between farms at random, but are chosen by farmers or allocated by the society according to some set of decision rules. Indeed it is likely that all inputs will vary together, as was found to be the case in the study of Chiweshe farmers just described. The skilled farmers use more land, labour, fertilizer, manure and fixed capital than the semi-skilled, who in turn use more of all these

resources than the unskilled. Where, as in this case, the levels of variable inputs are closely related between themselves, we speak of 'multicollinearity'. Its presence means that it is very difficult, if not impossible, to disentangle the influences of the variable inputs and obtain a reasonably precise estimate of their separate effects.

Take for instance the case of just two variable inputs, labour and land, which tend to vary together, more labour being employed on larger farms. It is then very difficult to say whether the larger output obtained on the larger farms is the result of the increased inputs of land or the increased inputs of labour, or how the extra output should be apportioned between the two inputs. We could only *safely* do this if there were some farmers who used extra land without using any more labour, that is if the inputs varied independently of each other and there was no multicollinearity present.

Where there is multicollinearity, it is particularly dangerous to omit one of the interrelated variable inputs from the function because then the marginal product of this input will be attributed to those left in the function. For example, imagine a situation where each extra hectare of land cultivated uses an extra unit of labour and yields an additional output of £26. Now this £26 represents the joint marginal product of both land and labour, but if land is left out of the production function it will appear that the £26 (or at least most of it) is the marginal product of labour alone. The omission of one of the interrelated variable inputs from the production function will therefore give over-estimates of the marginal productivities of the other inputs. It will give biased results. Because of this danger it is important that *all* variable inputs should be measured and included in the production function analysis. One input which is frequently omitted, because it is difficult to measure, is the input of management, but its omission gives rise to so-called 'management bias' in the estimated marginal products of the other resources used. In the Chiweshe study, management inputs are included but only at the three levels of skilled, semi-skilled and unskilled, and this ranking is based at least to some extent on the subjective judgement of the extension officers. If a more precise ranking of managerial inputs was possible, the estimates of marginal products for other inputs would probably be more accurate.

In summary it would appear that provided there is some independent variation between inputs, there is not 'exact multicollinearity' and it may be possible to estimate separate marginal products for all inputs. However, the reliability of the estimates may not be very good. Furthermore, it is important that all relevant variable inputs should be included in the function to avoid bias. It may be difficult to measure some inputs such as soil quality and management and large numbers of input variables may be involved. To reduce the number of variables some attempt at grouping inputs together may be necessary. For instance, child labour is grouped together with adult labour, different kinds of fixed capital are grouped together. Such grouping must involve some arbitrary choice of which variables to group together and what relative weightings to use.

Further reading

Dillon, J. L. (1977). *The Analysis of Response in Crop and Livestock Production*, 2nd edn, Oxford, Pergamon

Flinn, J. C. & Lagemann, J. (1980). Evaluating technical innovations under low resource farmer conditions, *Experimental Agriculture*, **16**, 91

Goldsworthy, P. R. (1967). Responses of cereals to fertilizers in Northern Nigeria: 1. Sorghum, *Experimental Agriculture*, **3**(1) 29

Idachaba, F. S. (1973). Marketing board crop taxation and input subsidies: a second-best approach, *Nigerian Journal of Economics and Social Studies* **15**, 317

Johnson, R. W. M. (1969). The African village economy: an analytical model, *The Farm Economist*, **11**(9) 359

Massell, B. F. (1967). Farm management in peasant agriculture: an empirical study, *Food Research Institute Studies*, **7**(2) 205

Norman, D. W., Hayward, J. A. & Hallam, H. R. (1975). Factors affecting cotton yields obtained by Nigerian farmers, *Cotton Growing Review*, **52**(1) 30

Norman, D. W., Pryor, D. H. & Gibbs, C. J. N. (1979). *Technical change and the small farmer in Hausaland, Northern Nigeria*. East Lansing, Michigan State University, Department of Agricultural Economics, African Rural Economy Paper 21

Osuntogun, A. (1978). The impact of co-operative credit on farm income and the efficiency of resource use in peasant agriculture: a case-study from three States in Nigeria, *African Journal of Agricultural Science*, **5**(2) 1

Saylor, R. G. (1974). Farm level cotton yields and the research and extension services in Tanzania, *Eastern Africa Journal of Rural Development* **7**, 46

Upton, M. (1979). The unproductive production function. *Journal of Agricultural Economics*, **30**(2) 179

Wonnacott, R. J. & Wonnacott, T. H. (1970). *Econometrics*, New York, Wiley

PART IV

Farm planning

14

Budgeting

The need for farm planning

Farm planning means assessing the implications of allocating resources in a particular way before deciding whether to act. It is an essential part of rational decision making and we have assumed in earlier chapters that farmers actually do plan what to produce and how to produce it. Thus planning is part of the day-to-day activity of running a farm; but it is particularly important when changes in the farm system are being considered. Few farmers would be willing to adopt an innovation without first evaluating the consequences.

However, for many farmers, planning is subjective and informal. Only on larger, commercial farms, where records and accounts are kept is it likely that formal procedures, such as budgeting are used. None the less, these practices are useful, not only in allowing closer control of the farm system, but also in persuading credit agents to provide loans. Thus there are good reasons for encouraging more farmers to keep records and accounts and to prepare budgets for proposed changes in their farming activities.

Farm planning techniques are also used by researchers, advisors and development planners, in two different ways. On the one hand, farming systems researchers and farm advisors are concerned to *prescribe* what farmers ought to do in order to advise on how systems can be improved. Development planners, on the other hand, may wish to *predict* how farmers will respond to changes in prices, institutions or technology. In either case the reliability of the results depends on how well the planner has identified farmer objectives and constraints. If the planner mistakenly assumes that the sole objective is profit maximization, his prescriptions may be unacceptable to farmers and his predictions will be wrong.

Farm planning techniques are designed for use on individual farms, taking account of the resource endowment, constraints and objectives peculiar to each household. However, given the large number of small farm households, the cost of providing individual advice and assistance would be prohibitive. A more feasible alternative is to prepare plans for one or more 'representative' case-study farms and to generalize from them to the whole target population. Thus Farming Systems Research involves assessing the impact of innovations throughout the recommendation domain, using plans for representative farm households. General guidelines for farmer advice and extension may be obtained in the same way. Similarly development planning may involve aggregation of budgets from representative case studies to predict the overall performance of a development project.

For planning purposes the 'representative' farm need not exist as a real farm. Plans can be drawn up for a model farm based on typical or average conditions. However, if the plans are then subjected to on-farm testing it is clearly necessary to identify actual case study or unit farms.

The budgeting procedure

A budget is simply an attempt to quantify the effects of a proposed plan. The prediction of inputs and outputs is an exercise in forecasting the future and there is obvious scope for error. Forecasts must, of course, be based on past experience under similar conditions. Budgets may be prepared at two different levels. Whole farm budgets are estimates of the overall impact of a proposed plan on the whole system. Partial budgets, on the other hand, are used where the plan only affects a particular enterprise or sub-system, which can then be considered separately. Only those items likely to change are included in the partial budget.

Budgeting involves three main steps:

(1) preparing a description and specification of the proposed plan, in terms of the area of each crop and number of each class of livestock to be produced and the methods of production;

(2) testing the feasibility of the proposed plan in terms of the resource requirements relative to what are available and to other institutional, social and cultural constraints;

(3) evaluation of the plan.

In budgeting, the specification of farm plans depends on the planner's intuitive judgement, although it must be influenced by technical knowledge of what crop varieties, types of livestock and methods of production are suited to the environment. Where it is part of a Farming Systems Research programme, however, it may be based on the diagnosis of farmer problems and limiting constraints. The specification is then designed to remove or overcome the constraints, where promising new technology is available. This stage is best achieved by active collaboration between the farm planner, crop and livestock specialists and, if possible, representative farmers.

The second stage involves estimation of whether sufficient land is available to allow the planned crops to be incorporated in a rotation which will maintain soil fertility. Seasonal labour requirements are compared with labour availability. Where appropriate, animal or machine power and cash constraints may also be considered. The plan may be tested to determine whether sufficient food will be produced to meet family needs.

The final evaluation involves assessing costs and benefits to the farm family. These are usually estimated in money terms, which may place undue emphasis on financial profits. However, as we have argued elsewhere, provided that other objectives are treated as constraints which must be satisfied, farmers may well prefer to choose the most profitable alternative. Nevertheless, if it is thought more appropriate, evaluation might be based on nutritional values for instance. Sensitivity analysis, in which the calculations are repeated for different values of key variables, may be used to give some guidance as to the risk involved.

Sources of planning data

In order to prepare a budget, estimates are needed of the resource input requirements, the yields, costs and benefits of each enterprise included in the plan. These are then compared with the resource base of the representative farm and used in evaluating the outcome. For ease of calculation it is assumed that each enterprise (a) uses inputs in fixed proportions and (b) is subject to constant returns to scale. This means that data on the average input and output per hectare of a crop can simply be multiplied by the number of hectares of that crop to arrive at the corresponding totals. Similarly, inputs and outputs per head of livestock may be multiplied by the number of animals. Planning data needs are thereby limited to information on the characteristics of the representative farm household plus average, per unit input and output for the relevant enterprises. These are known as 'input–output data', or 'coefficients'.

This approach does not necessarily imply that the law of diminishing returns and possibilities of input substitution have been ignored. Rather it is assumed that the least-cost combination of inputs has been found and is represented by the input–output coefficients used.

For crops and animals which have already been raised in the past, input–output data may be derived from farm records and accounts. Such information may have been collected during the description and diagnosis phase of Farming Systems Research. Where an innovation is proposed which has not previously been tested on farms, however, planning data must be derived from research experiments. Field studies have shown that farmers are rarely able to achieve the level of performance obtained on research stations. This is probably due to the much higher level of management that can be afforded on research stations. In any case, it means that the experimental results should be adjusted to allow for this difference before they are used for farm planning.

Input–output data for the main crop and livestock enterprises may be obtained from earlier studies and published sources, if more recent local data are not available. Indeed attempts have been made to provide standard, input–output data for general use, (eg see Phillips, 1964). Official agencies engaged in regular farm survey work, could provide a useful service in producing and updating standard planning data.

Crop mixtures raise special problems since it cannot be assumed that the total input or output for the mixture is simply the sum of the parts. The total monthly labour requirements of a mixture of yams and maize will differ from that of the two crops grown separately. Some operations such as clearing and weeding are shared while others such as harvesting are not. Thus there may be a case for treating the more common crop mixtures as single enterprises with input–output coefficients specific to them.

An example

The technique of whole farm budgeting will be illustrated with an example from Kenya (Clayton, 1961). The plan was devised by agricultural advisors for a representative smallholding of 4.30 hectares in the Star/Kikuyu grass ecological zone. It was intended that the proposed plan, if technically and economically feasible, would be recommended to other farmers in the area. Although prices and technical input–output coefficients may have changed since this study was reported, it is nevertheless useful in illustrating the principles.

The plan as specified is set out in Table 14.1. It consists essentially of coffee, English potatoes, maize and bean production, together with dairying for cash together with some subsistence production.

(i) *Land use* The first step in feasibility testing is to plan the pattern of land use. This involves comparing the area of each type of land required with the area available and determining sustainable crop rotations. In this case the area of coffee and of permanent grass had already been established, so the remaining task was to plan the allocation of arable land. It was further assumed that the area allocated to cassava, bananas and vegetables for home consumption should remain unchanged, as should the area of Napier grass for mulch and fodder. The land area remaining for arable crop rotation was 1.54 hectares.

Table 14.1. *Plan for smallholding in Kenya: Kagere Sub-location, Othaya Division, Nyeri District, Central Province.*
(a) Land use

	Hectares		Hectares
Homesteads and paths	0.26	Napier grass (fodder)	0.36
Arable rotation	1.54	Vegetables	
Coffee	0.40	(home consumption)	0.23
Napier grass (mulch)	0.42	Bananas	
Cassava (home consumption)	0.26	(home consumption)	0.23
Permanent grass and trees	0.60	Total	4.30

(b) Dairy cows up to maximum carrying capacity.

The choice of crop rotation, meaning the sequence of crops, the planting dates and the duration of each, also determines the area of each crop which can be grown. In this ecological zone there are two cropping seasons, the long rains and the short rains, so effectively the proposed seven year rotation involves 14 courses, as shown in Figure 14.1. To maintain a steady-state system, equal areas of land should be allocated initially to each 'year' of the rotation. This is, of course, assuming that the rotation allows fertility to be restored so the system can continue. Thus, in this case, the available area of 1.54 hectares should be divided into seven equal areas of 0.22 hectares. Of these, three are under grass initially, one is planted with beans, one with English potatoes and two with early maize. The resulting crop areas are

Grass ley	0.66 hectares
English potatoes	0.66 hectares
Beans	0.44 hectares
Early maize	0.44 hectares
Late maize	0.22 hectares

In this example the choice of rotation also determines the area of grazing available and hence the number of cows which can be carried.

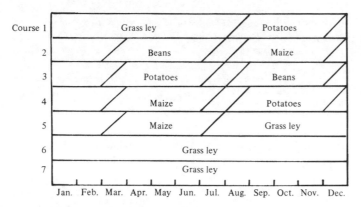

Fig. 14.1 The planned crop rotation

(ii) *Labour requirements* In comparing labour requirements of the plan with labour availability it is necessary to take account of the seasonality of farm work, the time needed for off-farm and household activities and possibly differences between men, women and children. Where there is strict division of labour, for instance, where men take sole responsibility for food crops, it may be necessary to treat male and female labour as separate categories. However, where work is shared, labour resources are pooled and should be treated as such. There may then be a case for using conversion factors to convert available female and child labour into standard man–days (see Chapter 7).

To allow for seasonality we need to estimate labour requirements and availability month by month (or week by week or at some other suitable interval). For this purpose standard monthly labour requirements per hectare of crops and per head of livestock are needed. The monthly total requirements can then be plotted on a graph as in Figure 7.1 to show the labour profile. From such a profile it is easy to identify where labour deficits and surpluses are likely to occur. This may suggest modifications of the plan to take up slack or to ease labour bottlenecks.

Preparation of a labour profile for the Kenyan farm showed that the plan was only feasible if oxen were hired for seed-bed cultivations, given a family work force of three full-time adult male equivalents. Assuming 300 working days annually, the monthly supply was estimated at 75 man–days. Although in no month did the needs of the plan exceed this, the total was approached in eight of the twelve months.

(iii) *Livestock feed* Here again, it is necessary to compare the livestock feed requirement, with the amount available, allowing for seasonal variations. Of course supplementary foods may be purchased but it is desirable to estimate the likely quantities and costs. Where, as in this example, animals are dependent on fodder grown on the farm, it may be necessary to adjust the numbers carried to match available supplies. Thus it was estimated that one cow required 0.4 hectares of grazing. Given a total of 1.62 hectares (0.66 of rotational ley, 0.36 of Napier grass and 0.60 of permanent grass), four cows could be carried.

In many situations, more complicated calculations would be needed. First if there were alternative competing classes of livestock, for instance cattle or sheep, or even different age categories it would be necessary to convert numbers into standard grazing livestock units for comparison with the carrying capacity of the grazed area. Second, if there were alternative feed sources available, supplies could only be aggregated in terms of nutritional measures such as MJ of metabolizable energy.

(iv) *Evaluation* Having estimated whether the proposed plan is feasible it may then be evaluated. The Kenyan farm plan was evaluated in money terms using estimates of the gross margin per hectare of crops and the gross margin per head of dairy cattle.

Table 14.2. *Estimated returns and costs in East African shillings* (s)

Enterprise	Number of units	Price of product	Yield	Gross output	Costs	Gross margin	Total gross margin
			per unit				
	hectares	s per bag	bags	s	s	s	s
English potatoes	0.66	13	98	1274	189	1085	716
Beans	0.44	43.25	15	648.75	32.75	616	271
Early maize	0.44	25	25	625	15	610	268
Late maize	0.22	25	20	500	14	486	107
			cherry parchment				
		s per kg	kg				
Coffee*	0.40	1.1	1253	9511	11	9500	3800
			per cow				
	cows	s per litre	litres	s	s	s	
Milk	4	0.33	910	300	50	250	1000
Total							6162
Less cost of oxen hire							140
Net income (return on capital, management and labour)							6022

*Parchment: cherry ratio 1:6.9.
After Clayton (1961).

The results are presented in Table 14.2. No charge was made for capital costs of investment in coffee trees, dairy cows and land improvements. Thus the estimated net income represents the overall cash return on land, labour and capital.

Comment

Unlike production function analysis, or linear programming to be discussed in Chapter 16, budgeting does not lead in a systematic way to an optimal, or most profitable, solution. It can be used to find the most profitable of a set of alternative plans but it does not guide the original choice of plans for evaluation. Production function analysis or linear programming, on the other hand, yields results in terms of the specific combinations of enterprises and levels of resource use which will yield the maximum profit.

None the less, budgeting will always have a place in practical farm planning because of its simplicity and flexibility. Budgets can be used to explore other aspects of farm plans than those discussed in our example. For instance, where irrigation is practised, water budgets may be used to assess seasonal crop water requirements for comparison with amounts available from precipitation and storage. Such analysis not only allows the feasibility of the proposed irrigation plan to be tested but also may suggest ways of economizing in water use.

Similarly, budgets may be used to assess seasonal requirements for animal draught power, machinery or equipment. As for labour, requirements and availability may be estimated in terms of working hours or days. Monthly cash flow budgets may be used to assess working-capital requirements. In this case it should be noted that revenues from sales of produce or off-farm earnings contribute to the supply of working capital, so the total supply is given by the cumulative *net* flow of cash into the household. Such an analysis will suggest where short-term credit may be needed, and whether borrowing can be justified.

Nutritional budgets are used to compare planned

150 *Farm planning*

Table 14.3. *Annual energy budget for an eight-cow Borana household*

Source	Gross energy value MJ per year
Energy directly produced	
Milk offtake 1802 l at 3.73 MJ	6 720
Slaughtered meat 84 kg liveweight at 5.34 MJ	449
Fallen meat (consumable mortalities) 140 kg at 4 MJ	560
Sub-total	7 729
Energy purchased	
Cereals 200 kg at 15 MJ	3 000
Sugar 31 kg at 14 MJ	434
Other foods	1 140
Sub-total	4 574
Total annual energy available	12 303
Basic annual energy requirement for three active adult male equivalents	12 000

(After Cossins & Upton 1986).

subsistence production with household food needs. Quantities are usually measured in MJ of gross energy, but checks may also be made as to whether sufficient protein or other essential nutrients will be available. Table 14.3 shows a simple energy budget for a typical Borana pastoralist household (Cossins & Upton, 1986).

Partial budgets

Where only a relatively minor change in the pattern of farming is proposed it is not necessary to prepare a complete farm budget to estimate the result. Instead, a partial budget can be used to arrive at the expected change in profits. This takes into account only those changes in costs and returns that result directly from the proposed modification. Farm costs or returns which are unaffected by the proposal are excluded from the calculation.

The simplest form of partial budget, applicable where a new enterprise or a new process such as the use of herbicides is introduced, involves the following questions:
(a) what extra returns (gains) can be expected?
(b) what extra costs will be incurred?

Where the proposed new activity substitutes for something already existing, as when one crop substitutes for another or a machine substitutes for labour, we must also ask:
(c) what present costs will no longer be incurred?
(d) what present income will be sacrificed?

Hence the total gain will be (a) + (c), the extra returns plus the saved costs, and the total cost will be (b) + (d), the extra costs plus the present income forgone. The total gain minus the total cost then represents the net gain or expected increase in profit.

As with complete budgets, the first step in partial budgeting should be a description of the proposed change stating clearly what is involved and when it occurs. Information should be included on the stock numbers, the areas of crop and the methods of production to be used. As a second step in partial budgeting it is useful to list those items in the existing system likely to be changed when the new policy is introduced. This reduces the likelihood of omitting possible indirect effects of the change. Appropriate values can then be used to predict extra returns, savings in present costs, extra costs and income sacrificed to arrive at the expected change in profit.

A partial budget can be compiled more quickly and easily than a complete budget, since it is only concerned with the costs and returns that are to be changed. The items of cost and income unaffected by the change need not be estimated.

There are obvious dangers in using partial budgeting since, as was argued in Chapter 10, even quite small changes in one enterprise may have repercussions throughout the whole farm system. The dangers are that some of these effects will be forgotten in making a partial analysis of the effects of the change. None the less, partial budgeting is generally preferred to whole farm budgeting, even for planning substantial changes, because fewer items must be estimated.

The example given in Table 14.4 is a partial budget for the adoption of a two-wheel tractor with a trailer and tool bar, injection planter and sprayer, developed at IITA and known as a 'farmobile'. For

Table 14.4. *Partial budget for farm mechanization*

	IITA farmobile 2-wheel tractor Dollars	Low horsepower 4-wheel tractor Dollars
Costs		
Fixed costs		
Depreciation	2 040	5 610
Interest	1 122	3 085
Operational costs	1 448	4 515
Variable costs of farm inputs	10 531	16 555
Total costs	15 141	29 765
Revenue from maize harvest	15 038	25 050
Revenue from cowpea harvest	7 555	12 705
Total revenue	22 593	37 755
Net benefit	7 452	7 990
Net benefit per hectare	857	555

(After IITA 1983).

comparison a partial budget for a small but conventional four-wheel tractor is included. These are clearly major items, likely to have a substantial impact on a smallholder farming system. Nonetheless, it is possible to leave other enterprises such as permanent crops and livestock out of the calculations and prepare partial budgets solely for the impact on arable crops. Those items left out of the budget are treated as being fixed. There is a possible advantage in that the results may be applicable to a wider range of different holdings than would a whole farm budget.

It may be noted that this example involves capital investment in machinery. The extra costs therefore include an allowance for depreciation. A similar approach is used in the partial budget for grain storage presented in Table 14.5. Although these analyses provide a useful assessment of the economic returns, it would, in theory, be more appropriate to use investment appraisal since capital investments are involved. We return to this topic in the next Chapter.

Table 14.5. *Partial budget to estimate extra profit from storage*

1 Specification
Proposal to store 10 tons (10 000 kg) sorghum per year. Cost of small concrete bin of 10 tons capacity £100. Expected life of bin ten years. No maintenance or repair cost. Grain to be fumigated, so losses due to insect damage and drying in the bin negligible.

2 Items in present system likely to be changed
Sorghum no longer available for sale at harvest time. No reduction in existing costs.

3 Estimated gains and cost (£)

Gains		Costs	
(a) Extra returns		(b) Extra costs	
Sales of stored grain		Cleaning bin	0.25
10 tons at £24.25	242.50	Fumigation of bin 15p per ton	1.50
		Depreciation of bin £100/10	10
(c) Saved costs	nil	(d) Present income sacrificed	
		Sales of grain at harvest-time (10 tons at £20)	200
Total	242.50		211.75

Net gain = 242.50 − 211.75 = 30.75

Fixed capital	£
Initial cost of bin	100
Working capital	
Income foregone by not selling sorghum at harvest-time	200
Cost of cleaning bin	0.25
Cost of fumigation	1.50
Total initial capital	301.75

Return on initial capital = 30.75/301.75 × 100 = 10%

Use of gross margins

In partial budgeting we are concerned only with those costs and returns which are expected to change. Other items, which are assumed to remain fixed are left out of the calculation. When the proposed change simply involves substituting one enterprise for another, the only costs that are likely to be affected are the variable or direct costs of each enterprise. Thus we may estimate the effect of substituting one hectare of cotton for one hectare of groundnuts by comparing the gross margins of the two crops. We are, of course, assuming that no extra land, labour or equipment is needed as a result of the change. Then

$$
\begin{aligned}
\text{net gain} &= (\text{extra returns} + \text{saved costs}) \\
&\quad - (\text{extra costs} + \text{income foregone}) \\
&= (\text{gross output of cotton} + \text{variable} \\
&\quad \text{costs of groundnuts}) \\
&\quad - (\text{variable costs of cotton} + \text{gross} \\
&\quad \text{output of groundnuts}) \\
&= (\text{gross margin of cotton}) - (\text{gross} \\
&\quad \text{margin of groundnuts}).
\end{aligned}
$$

This means that if we know the gross margins of the two enterprises it is a very quick and easy exercise to estimate the result of substituting a hectare of cotton for a hectare of groundnuts or even of substituting 0.6 hectares of cotton for 1.1 hectares of groundnuts or any other relative change that might be feasible.

Consideration of the feasibility of the change now suggests a way in which to plan on a more rational basis. If we can discover which resource is the most limiting constraint on production, we can select the enterprise which yields the highest gross margin per unit of that limiting constraint. Furthermore, we can decide objectively how far to expand that enterprise. In fact, we should expand it as far as possible until all the limiting resource is used up.

To illustrate let us assume, that the gross margins of cotton and groundnuts are £150 and £120 per hectare respectively and that labour is the resource which limits production. Hypothetical labour requirements per hectare of the two crops budgeted are set out in Table 14.6.

If labour is only available on a regular basis throughout the year, as is probably true of family labour, then the peak requirement must limit the area of crop that can be grown. In this case the peak requirement for both crops occurs in December, when cotton requires 25 man–days per hectare and groundnuts 20 man–days. Hence we should choose the crop which yields the highest gross margin per man–day of labour in December. For cotton the gross margin per hectare is £150 so the gross margin per man–day of December labour is £150/25 = £6. For groundnuts the comparable figure is £120/20 = £6. Hence there is nothing to choose between the two crops. However, if we assume that some of the

Table 14.6. *Monthly labour requirements per hectare of cotton and groundnuts*

	Man–days	
	Cotton	Groundnuts
January	10	14
February	8	—
March	—	—
April	12	—
May	22	—
June	24	16
July	17	12
August	17	8
September	—	7
October	—	8
November	15	15
December	25	20
Total	150	100

December work might be carried out in November, to spread the peak a little, then we should compare the gross margin per man–day required in November and December together. For cotton the total labour requirement for the two months is 40 man–days and for groundnuts it is 35. Now the gross margin per man–day over the two months is for cotton £150/40 = £3.75 and for groundnuts £120/35 = £3.43. Hence cotton is the more attractive alternative. The same is true if we take the December and January labour requirements together.

If we know just how much labour will be available in each month we can calculate the amount of cotton that may be grown. Let us say that a surplus of 20 man–days is available in each month for growing either cotton or groundnuts. If the December work load could not be spread into November then the maximum possible area of cotton would be 20/25 = 0.8 hectares. This would yield a gross margin of £150 × 0.8 = £120. However if we assume that it can be spread equally between the two months the labour available would be 40 man–days and the labour requirement per hectare 15 + 25 = 40. Hence a whole hectare could be grown, yielding a gross margin of £150.

This approach of selecting enterprises to introduce or expand according to the gross margin per unit of limiting resource is systematic and logical. So, too, is the policy of expanding that enterprise until the limiting resource is all used up.

However, in practice, several constraints may effectively limit the farmer's choice at the same time. A combination of enterprises may be needed to make best use of these limited resources as we saw in Chapter 4.

If the most limiting constraint could be clearly identified, a logical stepwise procedure might be adopted for planning based on (i) choosing the enterprise which yields the highest return to this constraint, (ii) observing which other constraints become effective (iii) introducing a second enterprise to make better use of the second limiting resource and so on. Such techniques are generally referred to as 'programme planning'.

Programme planning is tedious and difficult to use and although it has advantages over budgeting in moving in a logical sequence towards more profitable plans, the economic optimum will not necessarily be found. For this purpose the more formal techniques of linear programming are needed (see Chapter 16).

Further reading

Clayton, E. S. (1961). Economic and technical optima in peasant agriculture, *Journal of Agricultural Economics,* **14**(3) 337

Collinson, M. P. (1972). *Farm Management in Peasant Agriculture: A Handbook for Rural Development Planning in Africa*, New York, Praeger

Cossins, N. & Upton, M. (1986). *The Productivity and Potential of the Southern Rangelands of Ethiopia*. Addis Ababa, International Livestock Centre for Africa

Dalton, G. E. (1973). Adaption of farm management theory to the problems of the small-scale farmer in West Africa, in Ofori, T. M. (ed.) *Factors of Agricultural Growth in West Africa*. Proceedings of an International Conference, Legon, University of Ghana, Institute of Statistics, Social and Economic Research

Dillon, J. L. & Hardaker, J. B. (1980). *Farm Management Research for Small Farmer Development*, Rome, FAO Agricultural Services Bulletin No. 41

IITA (1983). *Annual Report for 1982*. Ibadan, Nigeria, International Institute of Tropical Agriculture

Phillips, T. A. (1964). *An Agricultural Notebook*, 2nd edn, Ikeja, Nigeria, Longmans

15

Investment appraisal

Capital investment and time horizons

Many technological improvements are embodied in new forms of durable capital. This clearly applies to machinery and equipment innovations and to new varieties of permanent crops and livestock. Land improvements in destumping, terracing or irrigation works also represent long-term investments. In all such cases an ordinary, single-period budget is inadequate for estimating resource needs and evaluating the investment.

Investment appraisal, or to use its other name, capital budgeting involves estimating resource inputs, costs and benefits over the whole lifetime of a medium- or long-term investment. Discounting techniques are then used to provide a single measure of the desirability of the investment. Costs and benefits are almost always measured in money terms, and this planning tool is most relevant to investments made for commercial gain. When a loan is sought from a formal credit agency, it is normal practice to submit a capital budget for the proposed investment in support of the application. Many credit banks require such a budget as proof of the viability of the investment before a loan is considered.

The purposes of investment appraisal are
  (i) assessment of cash requirements for funding the investment;
 (ii) provision of a financial plan to determine credit needs and the schedule of repayments;
(iii) evaluation of the investment, to determine whether it is financially viable and worth undertaking.

The method is closely similar to that of 'project appraisal' or 'cost–benefit analysis', used by economic planners to evaluate projects from the national point of view. The only difference lies in the prices used for evaluation. Whereas the national planner uses estimates of opportunity costs to the nation in 'economic analysis', or modifies these measures to allow for income distributional impact of the project in 'social analysis', we are here concerned with 'financial analysis' based mainly on market prices. (For further details of project appraisal see Gittinger, 1982.)

The three main stages of an investment appraisal are
  (i) specification of the investment plan, including its size, timing, productive life and sources of finance;
 (ii) estimation of the series of costs, and benefits, over the lifetime of the investment;
(iii) evaluation of the investment using discounting techniques.

In practice, the first two stages may overlap since the size and timing of the project is likely to be influenced by the estimated costs. As a specific example, consider a plan to introduce dairy cows on a farm. The number of cows that can be kept, that is the size of the dairy enterprise, must depend upon the fodder and cash requirements per cow, in relation to the amounts available. If resources are particularly scarce and credit lacking, cow numbers may be built up gradually over several years, rather than the whole herd being purchased together at the outset.

An early decision must be made as to the planning horizon, or length of life of the project. This is clearly linked with the choice of replacement policy; how long to keep each capital asset before its disposal and possible replacement. Let us assume, for present purposes, that a replacement policy has been chosen, and the economic 'optimum' life of each asset is known. In determining the planning horizon, three main cases may be distinguished.

Case 1 Short- or medium-term investments

Short-term investments in working capital such as seeds, fertilizers, stored produce or labour hire, and medium-term investment in machinery and equipment, generally have a clearly defined, finite life. Whilst working capital is replaced every year, machines and equipment last several years. Investment in pigs and poultry are sometimes treated under this heading. Not only is the livestock replaced each year, but also the housing may only last a limited number of years. In such a case we may treat the life of the investment as the planning horizon. It really does not matter whether the investment will be replaced, it is treated as a 'one-off' case.

Case II Continuous replacement/upkeep and infinite planning horizon

Long-term investments include such crops as cocoa, coffee, tea, oil-palm and rubber, breeding livestock and irrigation and other land improvements. In these cases a policy of continuous replacement or upkeep may be adopted to maintain a steady-state system. Thus if oil-palms have an estimated productive life of 30 years, then 1/30 of the area may be replanted annually. Similarly 20 per cent of the beef breeding herd may be replaced each year. Careful annual upkeep and repair of irrigation works and land improvements is intended to maintain productive capacity permanently. Following this approach of continuous maintenance of the original investment, the project should continue indefinitely; it has an infinite life.

In this case, once the project reaches maturity, the annual costs and benefits may be expected to remain constant indefinitely. Such a constant and infinite stream of costs or benefits is known as a perpetuity, and is very easily discounted to give the present value (V) as follows

$$V = A/i \qquad \text{where } A = \text{annual cost or benefit}$$
$$i = \text{discount rate.}$$

Thus, if the project is expected to reach maturity in year five from which point on it will earn an annual net benefit of £A, as a perpetuity, this infinite stream can be replaced by a net benefit of £V in year five. In this way an infinite planning horizon is reduced to a five-year horizon.

Case III Long-term investment with irregular replacement

For some long-term investments it may be inappropriate to assume a continuous replacement so net benefits do not form a perpetuity. It may then be necessary to decide arbitrarily on a planning horizon of say twenty or thirty years. Strictly speaking the value of the terminal assets, remaining at the end of the period, should be included in the stream of benefits. However the precise estimation of costs and benefits beyond about twenty years ahead is relatively unimportant. When discounting, even at fairly low rates, the more distant future costs and benefits have a relatively low weighting. For instance, when discounted at 10 per cent, £1 expected in twenty year's time is worth only 15p now. Thus it makes little difference whether a twenty-year or a thirty-year horizon is chosen.

Cash flow budgets

As in the case of simple budgeting, it is important to test the feasibility of the proposed plan by comparing expected resource requirements with resource availability. It may be appropriate to prepare budgets for land, labour, irrigation water, livestock feed or household diets. However, the development over time of resource requirements must now be considered. Several land use, labour or dietary budgets may now be needed for different stages of the proposed investment. Such feasibility testing may suggest modifications or improvements to the original plan.

The estimates used in any planning exercise are necessarily subject to uncertainty, which is greater the further ahead we try to plan. Decisions made now about productive activities several years hence are unlikely to be implemented in exactly the way, and with the precise results presently foreseen. However, since capital investment involves a commitment to a particular set of activities in the future, there is a strong case for attempting to predict the outcome as accurately as possible.

Detailed estimates of future resource inputs and product outputs, combined with forecasts of future prices are used in preparing the cash flow budget. This is the profile or sequence of cash flows (extra income minus extra costs) over the life of the investment project. They are usually estimated on an annual basis, although shorter intervals might be used. Thus when evaluating crop storage it might be more appropriate to work with quarterly or even monthly cash flows. However, since interest is generally charged at an annual rate, medium and long term investments are best divided into yearly intervals. Most investments start with negative cash flows associated with the costs of establishing the project. Positive cash flows come later as the annual returns increase (see Figure 15.1).

Although money values are used in estimating cash flows, non-marketed items are included. For instance, increases in home-consumed produce resulting from the planned investment should be included as benefits, while the opportunity costs of family labour or land form part of the total cost. The estimation of these opportunity costs raises particular problems; and may require whole farm budgeting (see below) or Linear Programming (see Chapter 16).

The essential feature of cash-flow budgeting is that costs and benefits are accounted for *at the time they are expected to occur* (see Chapter 12). Thus if the plan involves purchase of an ox plough in the third year of the project, the whole cost of the plough is subtracted from the cash flow for that year. If it is resold three years later, the cash flow is increased accordingly in the sixth year of the project. Naturally no depreciation allowance is made as this would involve double counting of the replacement cost.

We are, of course, only concerned with the *extra* costs and *extra* benefits associated with the planned investment, and the differences between them which measure the extra or incremental cash flows. Thus the incremental cash flow in a particular year may be measured either as extra benefit due to project minus extra cost or as cash flow with project minus cash flow without the project.

In discounting we implicitly assume that transactions occur at yearly intervals, rather than

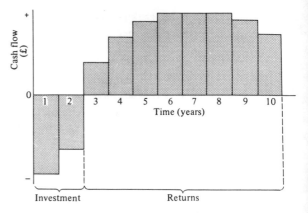

Fig. 15.1 Cash flows over time

continuously. Thus cost and benefits which arise in year 1 are assumed to occur at the end of the year. This means that the cash flow for year 1 is discounted one year in estimating the net present value. Similarly the cash flow for year 2 is discounted by two years and so on. The only exception to the rule that transactions are assumed to fall at the end of the year, is the initial investment. If the project is planned to start with the purchase of an item of capital, the purchase is assumed to occur at the *start* of year 1. To distinguish this date from the *end* of year 1, it is generally referred to as year 0.

Given that transactions are supposed to fall at the end of each year, it is necessary to make allowance for seasonal working capital requirements. If there is any increase in operating expenses between one year and the next, at least a part of the extra cost must be met in advance and this requires working capital. The allowance for additional working capital is included with the costs for the previous year in estimating cash flows. The ratio of working capital to operating costs depends upon the farming system. If only one crop is taken per year then the entire operating cost must be met before the crop is harvested. Working capital then represents 100 per cent of operating cost. If two crops are taken per year, only half the annual operating cost must be met before the first harvest, so working capital is around 50 per cent of operating cost (see Table 15.1). At the end of the project life, the total of these additional working capital requirements is added to the terminal benefits, to allow for the fact that it is recovered eventually.

Table 15.1. *Estimating working capital requirements*

	Year 1	Year 2	Year 3
Operating cost without working capital	£100	£150	£180
Operating cost with working capital, assuming two crops per year	£125	£165	

If credit is needed to finance the project, its impact may be measured by estimating cash flows after financing. This involves adding loan receipts to the cash flows for the relevant years and subtracting debt service payments. Two evaluations may then be made, one 'before financing' to give an overall assessment of the feasibility of the project and one 'after financing' to assess the effect on the farmer's income.

Where prices are rising due to inflation, the question arises of how to allow for this in investment analysis. Generally the answer is that we can ignore it so long as the effects of inflation are eliminated from the discount rate. This is justified on the grounds that all prices of inputs and outputs rise together at, say 15 per cent per year, if that is the inflation rate. But interest rates too will be increased by 15 per cent to cover inflation. Thus if we subtract 15 per cent from all interest and discount rates, and use constant prices the effects of inflation are eliminated. However, where credit is used, the repayments do not increase along with other costs under inflation. It is therefore necessary to discount loan repayments at the inflation rate to arrive at their estimated real cost.

An example

As an example we will consider investment in cocoa. The gross returns, costs and cash flows per hectare are given in Table 15.2. These values in Cedis are based on estimates made in the early 1970s for pure stand cocoa, at the lower yield limit. With higher cocoa yields or with inter-cropping the returns would be greater. Note that the initial negative cash flows are followed later by positive ones, and that from year 7 to year 30 the cash flows are equal (an annuity).

In Table 15.2 we also illustrate a possible disbursement and repayment schedule for a cocoa loan. The whole loan, sufficient to cover the establishment costs to the end of year 1 is paid at the outset. There is a five-year grace period, then the principal plus interest at 10 per cent are to be repaid by a series of 25 equal annual instalments. As a supplementary exercise, the reader may care to check that a 25-year annuity of 45 Cedis, discounted at 10 per cent is worth 408 Cedis. This, discounted a further five years for the grace period, has a present value of 254 Cedis which with a slight margin covers the loan. By adding in the loan, and subtracting the repayments when due, we arrive at the net cash flows after financing.

The discounting exercise is carried out in Table 15.3, assuming a discount rate of 18 per cent. Discount factors are taken from Appendix Table I. However, the annual flow of 195 Cedis from year 7 to year 30 is an annuity. Thus it must be multiplied by

Table 15.2. *Annual cash flows from cocoa* (Cedis per hectare)

	Year 0	Year 1	Year 2	Year 3	Year 4	Year 5	Year 6	Years 7 to 30
Gross returns	—	—	—	—	75	125	200	250
Costs	198	35	47	47	45	43	53	55
Cash flows	−198	−35	−47	−47	30	82	147	195
Loan disbursement and repayment	250	—	—	—	—	−45	−45	−45
Cash flows after financing	52	−35	−47	−47	30	37	102	150

(Adapted from Rourke 1974.)

Table 15.3. *Discounted cash flows from cocoa* (at 18 per cent)

Year number	Cash flow	Discount factor	Discounted cash flow	After financing	
				Cash flow	Discounted cash flow
0	−198	—	−198	52	52
1	−35	0.848	−30	−35	−30
2	−47	0.718	−34	−47	−34
3	−47	0.609	−29	−47	−29
4	30	0.516	15	30	15
5	82	0.437	36	37	16
6	147	0.370	54	102	38
7–30	195	0.314 × 5.432	333	150	256
Net present value			147		284

the annuity factor for 23 years at 18 per cent which is 5.432 (Appendix Table II).

The resultant present value of the annuity in year 7 is further multiplied by the discount factor for 7 years at 18 per cent, of 0.314 to give the present value at year zero. The sum of the discounted cash flows gives the Net Present Value (NPV) of 147 Cedis, per hectare of cocoa.

A similar discounting exercise is used to arrive at the NPV after financing. As might be expected, since the interest rate of the loan is only 10 per cent while the discount rate is 18 per cent, the NPV is increased by taking the loan.

We may also estimate the internal rate of return (IRR) for the investment (before financing). As suggested in Chapter 8, this involves a trial and error procedure. One approach is to estimate the NPV for various different discount rates and to plot the graph of NPV against the discount rate. This is shown in Figure 15.2 from which it is clear that the IRR (that is the discount rate where NPV is zero) is approximately 23 per cent.

Alternatively, given just two estimates of the NPV, one positive (at the lower discount rate) and one negative (at the higher discount rate) the following formula may be used (see Gittinger, 1982).

IRR = lower discount rate
  + Difference between discount rates
  × (NPV at lower discount rate/Sum of absolute values of NPVs at the two discount rates)

We have seen that the NPV at 18 per cent discount rate is 147 Cedis, (Table 15.3). At a discount rate of 24 per cent, the NPV is calculated as −21 Cedis. Hence the estimated IRR is found approximately to be

$$IRR = 18 + 6(147/168) = 23.25 \text{ per cent}$$

If we attempt to estimate the IRR after financing we run into problems. As we have seen the NPV at a discount rate of 18 per cent is 284 Cedis. Even at much higher discount rates the NPV is still positive. For instance at a discount rate of 50 per cent the NPV

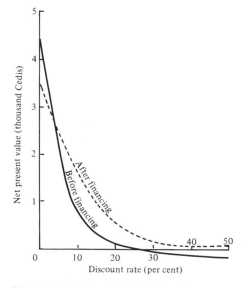

Fig. 15.2 The internal rate of return for cocoa

Table 15.4. *Farm level cash flow budget* (£)

| | Without project | With project | | | | |
		Year 0	Year 1	Year 2	Year 3	Years 4 to 25
Gross value of production	170	170	170	190	250	300
Incremental value of production		0	0	20	80	130
Operating expenses	20	220	50	50	50	50
Incremental operating expenses		200	30	30	30	30
Net benefit	150	−50	120	140	200	250
Incremental net benefit or cash flow		−200	−30	−10	50	100

is 31 Cedis, at 100 per cent it is 23 Cedis, at 200 per cent 34 Cedis and so on. In fact, if we plot NPV of the investment after financing against discount rate, as shown by the broken line in Figure 15.2, it is clear that the NPV never falls to zero. This means that no IRR exists.

Alternative measures

The possibility that an IRR may not exist is one reason why some authorities discourage the use of this measure in evaluating projects. However, for most standard projects of the type represented in Figure 15.1 where negative cash flows are followed by positive ones (ie there is only one change of sign in the sequence) an IRR does exist. It is particularly useful in comparing investments of different size and duration, where the NPV would be less meaningful.

There is another alternative approach to comparing investments of different duration, which is to estimate the equivalent annuity. This would enable us to compare different permanent crops. To illustrate, let us return to the cocoa crop (before financing) which at a discount rate of 18 per cent yields an NPV of 147 Cedis. This applies to a 30-year life. Now an annuity of one Cedi for 30 years discounted at 18 per cent gives a present value of 5.517 Cedis (see Appendix Table II). Hence the equivalent annual annuity, yielding an NPV of 147 Cedis is

147/5.517 = 26.64 Cedis per year.

This figure may now be compared with the return from annual crops, or the equivalent annual annuity for any other perennial crop.

Whole farm comparisons

The above example represented a form of partial budgeting. The cocoa enterprise was considered in isolation. However, in order to estimate cash flows it is necessary to assess the opportunity costs of all resources used. For the cocoa costing exercise, all labour both family and hired was valued at 1.00 Cedi per day, while land was assumed to have zero opportunity cost. Clearly, these are rather crude estimates of what the resources would earn in their best alternative use. An alternative approach is to determine the cost in terms of what income was earned without the project.

For this reason it is recommended that a form of whole farm budgeting should be used. The net benefits (cash flows before subtracting household resource costs) for the whole farm without the project should be subtracted from the net benefits for the whole farm with the project, to arrive at incremental net benefits, or cash flows.

The method is illustrated in Table 15.4, based on hypothetical figures for the improvements in production due to an Agricultural Development Project. For simplicity it is assumed that the net benefits without the project are constant from year to year, and that with the project they are constant from year 4 onwards. The project life is taken to be 25 years.

In the table we have subtracted the 'gross value of

production without project' from each of the annual gross values of production with the project, to arrive at the incremental values of each year. Similarly, we have estimated the incremental operating expenses for each year. These calculations are not essential for the estimation of the incremental net benefits or cash flows, since these can be obtained directly by subtracting 'net benefit without project' from each of the net benefits with the project. However, a cross check of the calculations may be useful. The incremental net benefits are then used in the discounting exercise to arrive at NPV or IRR.

Sensitivity analysis

For a given set of cash flows, discounted at a given rate, we obtain a single estimate of the NPV. In some unusual cases, more than one IRR may be found, or no IRR may be found as we have seen. However, in normal cases, a single estimate of the IRR is obtained. Clearly, when planning an investment for the future we cannot be sure that the precise estimate of NPV or IRR will be achieved. Some allowance should be made for risk.

The simplest method of allowing for the uncertainty associated with investment planning is sensitivity analysis. This involves:

(i) identifying key variables which first are likely to have a major impact on project performance and second are variable or uncertain;

(ii) repeating the discounting analysis for high and low values for each of the key variables.

As a result we obtain a set of estimates of the NPV or IRR. The whole exercise may prove a little tedious if done by hand, but it is very easily carried out by computer.

There are two possible weaknesses associated with sensitivity analysis. One is that the number of possible combinations of different levels for each of the key variables may prove very large. For instance if there are six key variables, each with high, average and low values, the number of possible combinations is

$$3^6 = 729$$

Possibly some of these combinations are so unlikely as to be not worth testing. The probability of achieving low yields, low prices, and high costs of inputs together may be very small indeed. However, since no attempt is made to estimate probabilities, the choice of which combinations of key variables to test becomes purely a matter of judgement. The analyst may judge that it is not worth considering the more favourable combinations of key variables. If the project looks viable on average, it will appear even more so when optimistic values are used for the key variables.

The other weakness is that sensitivity analysis gives no clear guidance as to whether a project should be accepted or not. For instance, suppose we have a project for which the expected NPV is positive but that after sensitivity analysis, eight out of twenty estimates of the NPV are negative. We have no clear guidance as to whether the project is acceptable. Despite these weaknesses, sensitivity analysis does provide some measure of the risk attached to a particular project. It also gives guidance as to where careful management will be needed because the final result is sensitive to the variable concerned. (For a more elaborate method of risk analysis, see Reutlinger, 1970.)

Further reading

Brown, M. L. (1979). *Farm Budgets: From Farm Income Analysis to Agricultural Project Analysis*, Baltimore, Maryland, Johns Hopkins

Gittinger, J. P. (1982). *Economic Analysis of Agricultural Projects*, 2nd edn, Baltimore, Maryland, Johns Hopkins

Reutlinger, S. (1970). *Techniques for Project Appraisal under Uncertainty*, Baltimore, Maryland, Johns Hopkins

Rourke, B. E. (1974). Profitability of cocoa and alternative crops in Eastern Region, Ghana. In *Economics of Cocoa Production and Marketing*, eds Kotey, R. A., Okali, C. & Rourke, B. E., University of Ghana, Legon, Institute of Statistical, Social and Economic Research

Upton, M. (1966). Tree crops: a long term investment, *Journal of Agricultural Economics*, **17**(1) 82

16

Linear programming

The assumptions

Linear programming (LP) is a systematic, mathematical procedure for finding the optimal plan, or programme, for a given set of conditions. To use this method the conditions must be presented in the following form:
(1) a limited choice of several activities;
(2) certain fixed constraints affecting the choice;
(3) straight line (linear) relationships.
These basic assumptions made in linear programming need further explanation.

The alternative activities may correspond with the crops and animals which could be produced on the farm being planned, but the word 'activity' as used in linear programming does not necessarily mean the same as the word 'enterprise'. Thus buying activities may be considered in drawing up the plan. Furthermore, several different activities might be associated with the same enterprise, where several alternative methods of production are possible. Yams grown in a mixture represent a different activity from yams grown alone; irrigated cotton represents a different activity from rain-fed cotton. In some cases it may be appropriate to treat a particular crop mixture as a single activity. The choice must be limited to a suitable number of activities for calculation. For the simple examples to be worked out in this Chapter, the number of activities is limited to two or three. However, larger problems only differ in the amount of arithmetic involved; the nature of the arithmetic is the same. For this reason most real linear programming problems are solved on computers which can deal with hundreds of alternative activities. Even so there is still a limit on the number of activities which can be considered.

The fixed constraints restrict the combinations of activities which are feasible. A plan which violates any constraint is assumed not to be feasible, which means that the constraint is assumed to be rigidly fixed. Thus, if July labour, limited to 28 man–days, is a constraint, then a plan requiring 30 man–days of July labour would be rejected as not feasible in linear programming, although in practice it might be possible to manage the extra work. The constraints may be physical quantities of productive resources, they may be technical requirements such as that cotton cannot be grown more than two years in direct succession, or they may be conditions which the farmer insists on for personal reasons. Many constraints are open-ended, which means that they need not be met exactly but are either maximum or minimum limits. For instance, although the total use of July labour cannot exceed 28 man–days, it may be less than this; the requirement for July labour must be equal to, or less than, the quantity available.

The assumption of linearity means that no matter how many units of a particular activity are included in the plan, the cost and return per unit remains the same. Thus if 1 hectare of maize needs 8 man–days of July labour and yields a gross margin of £100 then 3 hectares of maize are assumed to need $3 \times 8 = 24$ man–days of July labour and to yield a gross margin of $3 \times £100 = £300$. Likewise a quarter of an hectare is assumed to need $8/4 = 2$ man–days of July labour and to yield $£100/4 = £25$. The relationships between inputs and outputs are assumed to be straight lines. This can result in linear-programmed plans which include unrealistic fractions, particularly where livestock enterprises are involved. In a wide variety of problems the precision lost in rounding fractions to whole numbers is not sufficient to invalidate the

Table 16.1. *Data for linear programming problem*

| | Activities | | Constraint level |
	Maize	Groundnut	
Gross margin	£100	£320	
Constraints			
Land (hectares)	1	1	3
July labour (man–days)	8	16	28
November labour (man–days)	4	16	24
Total labour (man–days)	50	100	250

solution. However, many linear programming packages for the computer now include an option for specifying that certain variables can only take on integer, or whole-number, values. This 'integer programming' option is clearly an advance on ordinary LP but it does involve more computations. For the present we ignore this alternative.

The data needed for linear programming consist of a specification of the alternative activities to be considered and the return or gross margin per unit of each activity. The constraints must also be identified and the total capacity of each estimated, together with the demands of each activity on each constraint. These demands for inputs per unit of output are sometimes known as 'input–output coefficients'.

An example

We will take, as an example, a highly simplified farm planning problem with just two alternative activities, maize production and groundnut production, and four constraints. The objective is to maximize total gross margin. The constraints are land area, limited to 3 hectares, labour in the peak requirement months of July and November limited to 28 man–days and 24 man–days respectively, and total labour limited to 250 man–days. Each hectare of maize requires 8 man–days of labour in July, 4 man–days of labour in November and 50 man–days of labour in total and yields a gross margin of £100. Each hectare of groundnuts requires 16 man–days of labour in each of

the peak months and 100 man–days of labour in total to yield a gross margin of £320. These data are set out in Table 16.1.

If we use the symbols X_1 and X_2 to represent the number of units (hectares in this case) of maize and groundnuts in a plan and Z to represent the total gross margin, we can express the planning problem by a set of mathematical relationships as follows. Note that the symbol $\leq$ means 'is less than or equal to' and $\geq$ 'is greater than or equal to'.

Maximize total gross margin Z where

$$Z = 100X_1 + 320X_2 \qquad (1.1)$$

Subject to (ST) the constraints,

$$1X_1 + 1X_2 \leq 3 \text{ (hectares of land)}$$
$$8X_1 + 16X_2 \leq 28 \text{ (man–days of July labour)}$$
$$4X_1 + 16X_2 \leq 24 \text{ (man–days of November labour)} \qquad (1.2)$$
$$50X_1 + 100X_2 \leq 250 \text{ (man–days of total labour)}$$

and the non-negativity constraints

$$X_1 \geq 0, X_2 \geq 0. \qquad (1.3)$$

(1)

This is the standard form for a linear programming problem. It consists of three parts, (1.1) the function (eg profit or total gross margin) to be maximized, which is called 'the objective function', (1.2) the ordinary structural constraints and (1.3) the non-negative conditions on the variables. These need to be specified since a negative activity would have no acceptable meaning within the framework of the model. The structural constraints are expressed as inequalities, to allow for the possibility of leaving land fallow or of leaving labour unemployed at certain times of the year. It may be noted that if these had been expressed as exact equalities, implying that all available resources had to be used up in production, the problem would have *no* solution. The four constraint equations could not be satisfied simultaneously, with only two variables, X_1 and X_2. Thus we allow for non-use or disposal of the constraining resources.

This problem is a specific example of the general

LP maximizing model which may be written as follows:

$$\text{maximize } Z = \sum c_j \times x_j \qquad (2.1)$$
$$\sum a_{ij} \times x_j \leq b_i \qquad (2.2)$$
$$(i = 1 \text{ to } n)$$
$$\text{and} \qquad x_j \geq 0 \,(j = 1 \text{ to } m) \qquad (2.3)$$

$\left.\right\}$ (2)

or even more concisely using matrix and vector notation as;

$$\text{maximize } Z = \mathbf{c'x} \qquad (3.1)$$
$$\mathbf{Ax} \leq \mathbf{b} \qquad (3.2)$$
$$\mathbf{x} \geq \mathbf{0} \qquad (3.3)$$

$\left.\right\}$ (3)

where the x_j are the m activity levels,
  the c_j are the m per unit revenues or gross margins,
  the a_{ij} are the $(n \times m)$ input–output coefficients,
and the b_i are the n constraint levels.

Thus LP is simply a formal method of maximizing returns subject to a set of constraints. It deals with the economic problem of allocating scarce resources (constraints) among alternative competing uses (activities) to achieve desired ends (the objective function). Two methods of solution will be illustrated using our simple example. First a graphical method, based on plotting the production possibility boundary as in Chapter 4, is used. Then a more general, formal procedure, known as the 'simplex method' will be described.

A graphical approach

Feasible combinations of maize and groundnut production are plotted in Figure 16.1. Note that each of the constraint lines represents full use of the resource so there is none in disposal. For instance, consider the land constraint line AB. The equation of this line is $1X_1 + 1X_2 = 3$, which means that the area of maize plus the area of groundnuts is equal to the total area of 3 hectares. On this line, the land constraint is said to be 'binding' or 'effective'. Points below and to the left of this line, however, such as L or F, represent plans which leave some land in disposal, so this constraint is not effective.

Similarly, line CD represents combinations of the

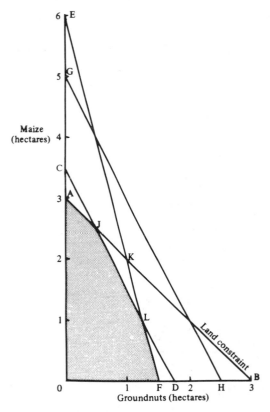

Fig. 16.1 The feasible region

two activities, for which the July labour constraint is binding, so none of this resource is in disposal. Line EF represents combinations of maize and groundnut production which would exhaust all the available labour in November, so there would be none of this resource in disposal, and the constraint would be binding. Finally, line GH represents the total labour constraint.

It should now be apparent that all feasible combinations of maize and groundnut production are represented by the area OAJLF, which is shaded. Points outside this area such as K, G or H are infeasible. Thus the total labour constraint line GH lies entirely outside the feasible area, so this constraint can never be effective. There will always be some labour in disposal or unemployed at certain times of the year. The total labour constraint is said to be 'dominated'

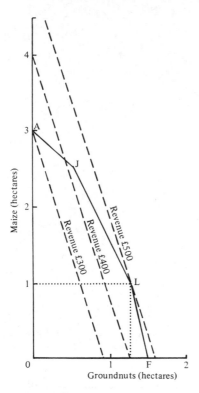

Fig. 16.2 The optimal solution

seeking an optimum solution we need only consider the combinations represented by extreme points of the feasible region. In this way LP greatly simplifies the problem of finding the optimum. Even in this very simple example with only two alternative activities of maize growing and groundnut growing, the number of feasible combinations is infinite. Any combination of the two, represented by a point inside the area OAJLF, would be feasible. Using simple budgeting we could estimate the total gross margin from even hundreds of these combinations and still never find the optimum, but the LP principle tells us we need only consider the extreme points of which, in this case, there are just five.

The same rule applies even when we have many activities and constraints so the graphical method cannot be used. At each extreme point we are, in effect, setting the number of real activities in the plan equal to the number of binding constraints. Thus the number of variables (activity levels) is equal to the number of equations (effective constraints) and an exact solution can be found. Such a solution is known as a 'basic feasible solution' (BFS). The fundamental principle of LP may therefore be restated as 'the optimum solution will always be found from among the basic feasible solutions'.

These ideas may be illustrated by Figures 16.1 and 16.2. Points A, J, L and F represent BFSs. The levels of each activity in each of these BFSs are given in Table 16.2. Other points such as G and K represent basic solutions but of course they are not feasible. Point A represents one activity, maize production, and one binding constraint, land. Points J and L represent combinations of the two activities with two effective constraints, land and July labour at J, and July and November labour at L. Point F again represents only one activity, groundnut production and one constraint, November labour. As we have seen, the optimum occurs at one of these BFSs, namely point L.

The fact that, in the optimum solution, the number of real activities equals the number of effective constraints is of some practical importance. Although a farmer may be faced with many different constraints, if he is limited to say five alternative activities, he will only be able to satisfy five constraints. Other scarce resources will have to remain unused. Conversely, if

by other constraints and it could be omitted from the rest of the analysis.

Figure 16.2 shows the production possibility boundary from Figure 16.1, with isorevenue lines for total gross margins (Z) of £300, £400 and £500. Clearly the maximum feasible gross margin is £500 obtained from a combination of 1.25 hectares of groundnuts and 1 hectare of maize (point L). This then is the optimum solution.

Some useful general conclusions may be drawn from further consideration of this illustration. First, it may be noted that, since the aim is to produce more rather than less, the optimal solution must always lie on the boundary of the feasible area; no matter what the relative product prices or gross margins are. Secondly, because the boundary is made up of straight line segments, the optimum must always occur at a corner, which is an extreme point of the feasible area.

This important result represents the fundamental principle of linear programming (LP), namely that in

Table 16.2. *The basic feasible solutions*

| Extreme point | Real activities | | Amount of resource unused (man–days) | | | | Total gross margin |
	Maize X_1	Groundnut X_2	Land (ha)	July labour	November labour	Total labour	
0	0	0	3	28	24	250	0
A	3	0	0	4	12	100	300
J	$2\frac{1}{2}$	$\frac{1}{2}$	0	0	6	75	410
L	1	$1\frac{1}{4}$	$\frac{3}{4}$	0	0	75	500
F	0	$1\frac{1}{2}$	$1\frac{1}{2}$	4	0	100	480

there are few critical constraints, few activities are needed and a simple system will be adequate. The apparent complexity of many African farms may reflect an attempt to make best use of a multitude of different constraints.

Shadow prices

For each BFS of an LP problem there is a corresponding set of shadow prices. These are the opportunity costs of the resources used, and reflect their relative scarcity. For a resource which is in disposal, and the constraint is not effective, the shadow price is zero. There is no opportunity cost of using an extra unit of such a resource because there is a surplus already available. For resources which are effective constraints the opportunity cost is the revenue foregone when one unit of the resource is released. It represents the value of the marginal product, and this in turn is the maximum amount it is worth paying to hire in an extra unit of the resource. In summary, there are three alternative views of a shadow price:

(i) the opportunity cost per unit
(ii) the marginal value product per unit
(iii) the maximum amount it is worth paying to hire the resource.

Again the shadow price may be illustrated using our example. At point A, where 3 hectares of maize are produced land is the only effective constraint. There is surplus labour at all periods, so the shadow price of labour is zero. However, each hectare of maize yields a return of £100, so the cost of releasing one unit of land is £100. This is the shadow price. At point F, November labour is the only effective con-straint. Each man–day used in growing groundnuts produces £320/16 = £20. Hence this is the shadow price of November labour.

The calculation of shadow prices is more complicated where there is more than one effective constraint. Let us consider the optimum solution at point L, representing a total revenue of £500. Here there are two effective constraints July labour and November labour and we need to separate their effects. Consider first, a reduction in July labour from 28 to 24 man–days. This would cause the optimum solution to move from point L to point F, as the reader may check from Figure 16.2. Note that the line segment JL represents the July labour constraint, and that the assumed reduction would cause this line to move to a parallel position through A to F. Hence the optimum would move to F, which represents 1.5 hectares of groundnuts, yielding £320 × 1.5 = £480. The reduction of July labour by 4 man–days results in a reduction of total revenue by £20 (£500 − £480). The shadow price per man–day of July labour is therefore £20/4 = £5.

To calculate the shadow price of November labour, it is somewhat easier to assume an *increase* in availability to 28 man–days. This would then allow the optimum solution to move to point D, representing 1.75 hectares of groundnuts. The total revenue would then be £320 × 1.75 = £560. Thus an increase of 4 man–days of November labour allows an increase in revenue of £60 (£560 − £500). The shadow price per man–day of November labour is therefore £60/4 = £15. Note that this is different from the shadow price at point F. Land and total labour are in disposal so their shadow price is zero. We could calculate the shadow price of land and July labour at point J in a similar fashion.

Two further important points should be noted in relation to shadow prices. The first is that if the cost of each unit of resource inputs is assumed to be the same as the shadow price, the total revenue is exactly equal to the total cost. There is then no profit or loss. This applies to both the whole plan and to individual activities included in the plan.

For instance at point F, 1.5 hectares of groundnuts yield a total return of £480. At this point the shadow price of November labour is £20 for 24 man–days whilst it is zero for all other resources. Thus the total cost is £20 × 24 = £480, which is equal to total return. Clearly, the same result obtains if calculated on a per hectare basis.

At point L, the results are as follows:

	July labour cost		November labour cost		Total revenue
Whole plan per hectare	(£5 × 28)	+	(£15 × 24)	=	£500
maize per hectare	(£5 × 8)	+	(£15 × 4)	=	£100 (4)
groundnuts	(£5 × 16)	+	(£15 × 16)	=	£320

In fact, this provides another way of calculating shadow prices. They are those values for limited resources which account for the total return from activities in the plan exactly. The determination of shadow prices is sometimes described as the 'dual problem' in relation to the 'primal problem' which concerns the determination of an optimal plan. In practice, a knowledge of shadow prices may be as useful as or even more useful than the estimated best plan.

The second important point to note is that shadow prices obey the law of diminishing marginal returns. To illustrate, let us assume that inputs of July labour are gradually increased, whilst the other resources are held fixed at their constraint levels. The effect on total revenue (gross margin) is shown in Figure 16.3. The first 24 man–days of July labour may be used in groundnut production which yields the highest marginal value product of £20, per man–day. This takes us

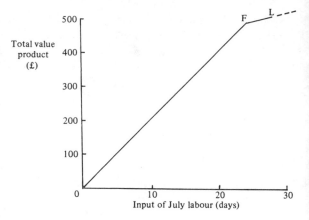

Fig. 16.3 Diminishing marginal returns in linear programming

to point F on Figure 16.2. Other constraints prevent the expansion of groundnut production beyond this point.

However, total revenue can be increased still further and more July labour can be employed by introducing maize production into the plan. Between point F and point L in Figure 16.2, some maize is substituted for groundnuts. As we have seen, the marginal value product (shadow price) is now £15. This decline of the marginal (value) product of July labour from £20 to £15 therefore reflects the operation of the law of diminishing marginal returns. None the less, if all inputs are increased together, that is to scale, there are constant returns because of the linearity assumption.

The simplex method

We turn now to a formal mathematical approach to linear programming. By following the procedures described here it should be possible to solve larger, more complicated problems than our present example. In principle, the method is applicable to problems with any numbers of activities and constraints, and forms the basis of most computer programmes for LP such as LINDO.

All formal methods of LP, involve a step-by-step approach which moves from one basic feasible solution to another; each move being known as an 'iteration'. It is not normally possible to move directly to an optimum solution in one iteration, except in the

Table 16.3. *Simplex tableau: Tableau I*

	Z	X_1	X_2	S_1	S_2	S_3	S_4	Constant b
Row 0	1	−100	−320	0	0	0	0	0
Row 1	0	1	1	1	0	0	0	3
Row 2	0	8	16	0	1	0	0	28
Row 3	0	4	16	0	0	1	0	24
Row 4	0	50	100	0	0	0	1	250

very simplest cases. The idea of the simplex method is to start from an initial BFS and then calculate whether the value of the objective function can be increased by moving to a neighbouring BFS. If so, the move is made (an iteration is completed) and again calculations are applied to determine whether further improvement is possible by a subsequent move. When finally a BFS is attained that does not admit of further improvement, this constitutes the optimum solution.

Thus there are two key features of the method:
(i) the calculations lead from one BFS to a neighbouring BFS at each iteration;
(ii) each movement is in the direction which increases the value of the objective function.

It may be viewed as a 'hill-climbing technique' for scaling the hill of increasing returns. If there is an optimum, it must eventually be reached. The method is efficient in that only a limited subset of BFSs are calculated in moving to the optimum. We shall illustrate the technique using the example from Table 16.1.

There are three preliminary steps to be carried out which will make the calculations easier.

Preliminary step 1: Introduce slack variables

Slack variables are used to represent the non-use or disposal of resources. Their use enables us to convert the constraint inequalities of the LP problem into equalities. In our example we will introduce four slack variables S_1 to S_4 to allow for disposal of each of the constraining variables. Variable S_1 represents non-use or fallowing land. It is measured in hectares, adds nothing to total gross margin and requires no labour. Similarly S_2, S_3 and S_4 represent leisure or non-use of labour and are measured in man–days.

Our illustrative example, equations (1), may now be written as follows:

Maximize:

$$Z = 100X_1 + 320X_2 + 0S_1 + 0S_2 + 0S_3 \\ + 0S_4 \qquad (5.1)$$

S.T.

$$\left. \begin{array}{l} 1X_1 + 1X_2 + 1S_1 + 0S_2 + 0S_3 + 0S_4 \ \ = 3 \\ 8X_1 + 16X_2 + 0S_1 + 1S_2 + 0S_3 + 0S_4 \ = 28 \quad (5.2) \\ 4X_1 + 16X_2 + 0S_1 + 0S_2 + 1S_3 + 0S_4 \ = 24 \\ 50X_1 + 100X_2 + 0S_1 + 0S_2 + 0S_3 + 1S_4 = 250 \end{array} \right\} (5)$$

and

$$X_1 \geqslant 0, X_2 \geqslant 0, S_1 \geqslant 0, S_2 \geqslant 0, S_3 \geqslant 0, S_4 \geqslant 0 \quad (5.3)$$

The zeros are included to complete the rectangular matrix of values.

Preliminary step 2: Rearrange the objective function

Although the purpose may not be immediately obvious, it is convenient to treat Z as yet another variable in the analysis and to move the right-hand side of the objective function over to the left as follows:

$$Z = 100X_1 + 320X_2 + 0S_1 + 0S_2 + 0S_3 + 0S_4$$

implies

$$Z - 100X_1 - 320X_2 - 0S_1 - 0S_2 - 0S_3 - 0S_4 = 0$$

Of course, the sign of the zero terms does not really matter; zero is neither positive nor negative.

Preliminary step 3: Set up the simplex tableau

This simply involves rearranging the data of equations (4) in the tidier form of a tableau, as shown in Table 16.3. A column is reserved for each variable, for real activities and slacks. In addition there is a column for Z and another on the right for the b-values or constraint levels. The corresponding coefficients are set out in the rows below, representing the original equations. The vertical line between the S_4 column and the b column is where the equal sign should be in the various equations. It is advisable to compare Table 16.3 with Table 16.1 to see how one is derived from the other.

It should be noted, that this Simplex tableau, actually represents a basic feasible solution, albeit a rather unattractive one, since the value of the objective function is zero. This particular solution is to produce nothing and leave all the resources in disposal. It may be read-off from the tableau by concentrating on those columns which consist only of ones and zeros. They are known as 'identity vectors'. These vectors form the, so-called 'basis' and from them we can identify the corresponding BFS. The X_1 and X_2 columns are not in the current basis and may be ignored temporarily. In effect we assume $X_1 = X_2 = 0$. The BFS is thus

$$
\begin{aligned}
Z &= 0 \\
S_1 &= 3 \\
S_2 &= 28 \\
S_3 &= 24 \\
S_4 &= 250
\end{aligned}
$$

The numbers in row 0, represent the costs, per unit, of introducing non-basic variables into the basis. As they are negative in this case, it means the costs are negative: that is there are positive gains to be made by introducing maize or groundnuts into the plan.

We are now ready to start the iterations which will lead to the optimum solution.

The iterations

Essentially, each iteration consists in substituting one activity from outside the basis for one that is already in the basis. Thus in the first iteration we will substitute a real activity for a disposal activity. The incoming activity is identified with a particular column, called the 'pivot column'. Each basis activity is associated with a particular row; for instance S_1 with Row 1, S_2 with Row 2 and so on in Tableau I. Thus the outgoing activity is associated with a particular 'pivot row'. The element which occurs where the selected column and row intersect is known as the

'pivot'. We can now describe the calculations involved in each iteration by a series of simple rules.

Rule 1. Choice of pivot column Scan Row 0, and select the element with the highest negative value. The corresponding column is the pivot column. This rule ensures that the incoming activity is the one which adds most to the value of the objective function. In Table 16.3 the pivot column is the one for groundnuts (X_2) since 320 is the largest negative value in Row 0.

Rule 2. Choice of pivot row For this purpose, we must divide each (positive) element in the pivot column into the corresponding element in the b column. The element for which the resulting quotient is smallest is in fact the pivot. (For our example these ratios are

Row 1 3/1   = 3
Row 2 28/16   = 1.75
Row 3 24/16   = 1.5
Row 4 250/100 = 2.5

Clearly the quotient 1.5 in Row 3 is the smallest, so this is the *pivot row*, and the number 16 in Row 3 of the groundnut activity column is the pivot. It is underlined in Table 16.3.)

The objective of this procedure is to ensure that the next basic solution will be feasible. By choosing the smallest ratio we ensure that all the constraints can be satisfied.

Rule 3. Derive the new pivot row This is achieved by dividing every element in the row by the pivot element. In our example this means dividing each element in Row 3 by 16, to give the results shown in Row 3 of the new tableau in Table 16.4.

Rule 4. Derive other rows The objective here is to replace all elements in the pivot column, other than the pivot itself, with zeros. This is achieved by subtracting (or adding) a suitable multiple of the pivot from (or to) the old element. Thus if the old element is 'a' then 'a/b' times 'b' must be subtracted from 'a' to yield zero.

However, to be consistent, the same operation must be applied to every element of the row. Thus each element of the pivot row must be multiplied by

Table 16.4. *Tableau II*

	Z	X_1	X_2	S_1	S_2	S_3	S_4	Constant b	R
Row 0	1	-20	0	0	0	20	0	480	
Row 1	0	$\frac{3}{4}$	0	1	0	$-\frac{1}{16}$	0	$1\frac{1}{2}$	2
Row 2	0	4	0	0	1	-1	0	4	1
Row 3	0	$\frac{1}{4}$	1	0	0	$\frac{1}{16}$	0	$1\frac{1}{2}$	6
Row 4	0	25	0	0	0	$-6\frac{1}{4}$	1	100	4

'a/b' before subtracting it from the corresponding element of the old 'a' row.

The elements of the new Row 0 shown in Table 16.4 are derived as follows. Note first that to reduce -320 to zero requires *adding* $320/16 = 20$ times the pivot element. Thus the whole Row 0 becomes

$$1, -100 + (20 \times 4), -320 + (20 \times 16), 0, 0, + (20 \times 1), 0, + (20 \times 24)$$

Similarly, the new Row 1 is obtained from the old Row 1 by subtracting 1/16 times the pivot row to give

$$0, 1 - 4/16, 1 - 16/16, 1, 0, -1/16, 0, 3 - 24/16.$$

The results of similar calculations for each of the rows are given in the new tableau of Table 16.4.

Rule 5. Inspect the new tableau In particular Row 0 should be scanned. If there are no negative elements, the optimum has been found. If on the other hand there are negative elements, then a further iteration is required, for which we return to rule 1 and repeat the cycle. However, the calculations are now all based on the new tableau. The process is repeated until an optimum is found.

From Table 16.4, Tableau II we see that the result of the first iteration is a plan to grow $1\frac{1}{2}$ hectares of groundnuts. From the set of identity vectors we see that the whole plan is

$Z\ = 480$ (total gross margin)
$X_2 = 1\frac{1}{2}$ (hectares of groundnuts)
$S_1 = 1\frac{1}{2}$ (hectares of land unused)
$S_2 = 4$ (man–days of July labour unused)
$S_4 = 100$ (man–days of total labour unused)

However, it is clear that this plan can be improved upon by introducing the maize growing activity X_1,

(coefficient of -20 in Row 0). In Table 16.4 an extra R column has been added to show the outcome of dividing elements of the X_1 column into the corresponding elements of the b column. Clearly, Row 2 is the pivot row, and unused July labour S_2 leaves the basis. In other words July labour will be fully employed.

The results of the second iteration are shown in Table 16.5; Tableau III. It is clear that this is the optimal solution since no negative values appear in Row 0. As we have already seen it represents 1 hectare of maize and $1\frac{1}{4}$ hectares of groundnuts. This leaves $\frac{3}{4}$ hectare of land in disposal and 75 man–days of labour unused over the whole year.

The BFSs represented by each of these three tableaux are represented by extreme points in Figure 16.2. Thus the origin represents Tableau I, point F represents Tableau II and point L, the optimum, represents Tableau III. The Simplex method has allowed us to find an optimum without even considering the BFSs represented by points A and J.

Interpreting a tableau

As we have seen, the identity vectors, together with the right-hand side (**b**) vector, allow us to read off the current farm plan at each iteration. At the same time Row 0 gives the current values of the shadow prices. Thus the Simplex method provides both a primal and a dual solution for each basis.

The two are optimized together. This means that the *true* opportunity costs of scarce resources are given by the shadow prices at the optimal solution.

The figures in the remaining columns, for variables not in the basis, are the rates at which they substitute for activities which *are* in the basis. For instance the

Table 16.5. *Tableau III: The optimum*

	Z	X_1	X_2	S_1	S_2	S_3	S_4	Constant b
Row 0	1	0	0	0	5	15	0	500
Row 1	0	0	0	1	$-\frac{3}{16}$	$\frac{1}{8}$	0	$\frac{3}{4}$
Row 2	0	1	0	0	$\frac{1}{4}$	$-\frac{1}{4}$	0	1
Row 3	0	0	1	0	$-\frac{1}{16}$	$\frac{1}{8}$	0	$1\frac{1}{4}$
Row 4	0	0	0	0	$-6\frac{1}{4}$	0	1	75

(S_3) November labour column in Table 16.5, tells us first that the shadow price is £15. This arises because one man–day of November substitutes for $\frac{1}{8}$ hectare of land (row 1) and $\frac{1}{8}$ hectare of groundnuts (row 3). However, it would result in an *increase* of $\frac{1}{4}$ hectare of maize (row 2) shown by a negative rate of substitution. In fact, these numbers are incidental to the main calculations and need not concern us further.

From Row 0 of Table 16.5 we see that shadow prices for the four resource constraints are zero for land, £5 per man–day for July labour, £15 per man–day for November labour and zero for total labour. These shadow prices can be used for partial budgeting to estimate the desirability of introducing a new activity on to the farm. For instance, suppose our farmer could grow cotton as an alternative to maize and groundnuts. The gross margin for cotton is £400 per hectare but the labour requirements are 32 man–days and 20 man–days in July and November respectively. We can calculate the net cost of introducing this crop into the farm plan as follows:

> Opportunity costs of resources minus revenue equals net cost
> £(32 × 5) + (20 × 15) − 400 = £60
> July labour   November labour

Clearly, although cotton returns a higher gross margin per hectare than either maize or groundnuts, the high labour cost makes cotton an unattractive alternative. Total farm gross margin would be reduced by £60 for each hectare of cotton introduced.

Readers may find it instructive to include this cotton growing activity in the initial tableau and to repeat the calculations to find the optimal solution. Although cotton will enter the basis at the outset because of the high return per hectare, it will then be forced out again because of the high labour requirement. The optimal solution will be no different from that shown in Table 16.5. Cotton production will end up as a non-basis activity with a 'shadow price' of £60 which represents the cost of introducing one hectare of the activity into the optimal plan.

In fact, a general conclusion may be drawn: of the alternative productive activities considered, those which are included in the optimal plan have zero net cost (cost of resources valued at their shadow prices exactly equals revenue as shown in equations (4)), while those which are excluded have a positive net cost. This mirrors the result for resource constraints, that if some of the resource is in disposal the shadow price is zero, whereas if the constraint is effective a positive shadow price applies. Note that negative shadow prices must imply that the optimum has not been found. There are rare cases, of so-called 'degeneracy', in which these conclusions do not apply, but otherwise they are generally true.

Some extensions

LP is a much more flexible tool than is apparent from the discussion so far. In practice, many more complicated relationships may be modelled, but skill is needed in building the initial Tableau. Some useful extensions will now be outlined.

(i) Parametric programming

This is simply ordinary LP in which either (a) one or more of the values in the objective function or (b) one or more of the resource constraints are allowed to vary. The effects of these changes on the optimal plan can then be estimated. Many computer programmes for LP allow for these options within a single programming run. However, the range of alternatives compared can be extended to include new activities or changes in the input–output coefficients simply by rerunning the programme.

Parametric programming may be used in various ways:

(a) sensitivity analysis, to test the stability of the optimal solution in view of possible errors in estimating the data used;

(b) risk analysis, to test the effect of chance variations in the data (eg see Heyer, 1972; Farrington, 1976);

(c) prediction of farmer supply (or input demand) response to changing prices. This of course depends on the assumptions that farmers are maximizers and that the model farm is representative of the whole population of farms. (See Ogunfowora, 1972, or Ogunfowora & Norman, 1973.)

(d) evaluation of the impact of alternative technologies on farming systems and farmer incomes. Again the model farm is assumed to represent a larger population while coefficients for new technologies are based on research station trials. This application of LP clearly relates to farming systems research. Mixed cropping (Ogunfowora & Norman, 1974), zero tillage systems (Knipscheer, Menz & Verinumbe, 1983) and mechanization practices (Ahmed & Kinsey, 1984) have been evaluated in this way.

(ii) *Buying activities and tie rows*

The purchase or hire of inputs is easily incorporated in an LP problem. Each buying activity has a positive cost in row 0 and contributes to the relevant constraints row. It therefore has a negative coefficient. Thus we might incorporate a labour hiring activity, at £2 per day, for the month of July (X_3) into our model of Table 16.3 by adding the following column of coefficients.

row 0	Cost	2
row 1	Land	0
row 2	July labour	−1
row 3	November labour	0
row 4	Total labour	−1

Note that labour hire in July also contributes to total annual labour availability. The effect on the optimal plan is as follows, as the reader may readily check by repeating the analysis.

Total gross margin: Z	= £520
Labour hire in July: X_3	= 4 man–days
Maize production: X_1	= 2 hectares
Groundnut production: X_2	= 1 hectare
Total labour in disposal: S_4	= 54 man–days

The shadow prices are now

Land, £16; July labour, £2; November labour, £17.

In some cases an extra constraint should be added to limit the level of hiring or buying. A limit may be imposed by shortage of funds for instance.

In this example, the July labour constraint row becomes a 'tie-row'; since it ties the labour hiring activity to the labour-using activities of maize and groundnut production. Indeed, a situation might arise in which no family labour was available in July, so labour would have to be hired. The tie row would then read

$$8X_1 + 16X_2 - 1X_3 + 1S_2 = 0$$

with all variables as already defined. Thus the labour hiring activity (X_3) yields a 'permit' (represented by the negative coefficient) to introduce other labour-using activities.

A similar technique may be used to tie a bush clearing activity to subsequent cropping activities, through a tie row for cleared land. A livestock producing activity may be tied to a fodder-producing activity through the fodder tie row. Furthermore rotational constraints, on the maximum proportion devoted to certain crops, may be imposed in the same way. For instance, in an early LP study of mixed farming in the Nyeri district of Kenya, a constraint was imposed which required that at least four years of grass ley are included for each three years of cropping, (Clayton, 1961). The rotational tie-line is

$$3X_1 + 3X_2 + 3X_3 - 4X_4 \leqslant 0$$

where X_1, X_2 and X_3 are hectares of arable-crop activities and X_4 represents hectares of ley.

Table 16.6. *Finding an initial BFS*

Tableau I

		Z	X_1	X_2	S_1	S_2	S_3	S_4	S_5	Q_1	Constant b
0	Objective	1	−100	−320						1000	
1	Land		1	1	1						3
2	July labour		8	16		1					28
3	November labour		4	16			1				24
4	Total labour		50	100				1			250
5	Diet		7.5	5					−1	1	15

Tableau II

		Z	X_1	X_2	S_1	S_2	S_3	S_4	S_5	Q_1	Constant b
0	Objective	1	−7600	−5320					1000	0	−15 000
1	Land		1	1	1						3
2	July labour		8	16		1					28
3	November labour		4	16			1				24
4	Total labour		50	100				1			250
5	Diet		7.5	5					−1	1	15

(iii) *Minimum target constraints*

Let us now return to the original problem of Table 16.1, but introduce the additional constraint that the farmer needs to produce, *at least* 15 000 MJ of food energy for family subsistence. Maize yields 7500 MJ per hectare and groundnuts 5000 MJ per hectare. This dietary constraint may be written in GJ (1GJ = 1000 MJ) as follows:

$$7.5X_1 + 5X_2 \geqslant 15$$

Note that the inequality is the reverse of those dealt with so far; total energy production may exceed 15 GJ but must not fall below this level. If a slack variable (S_5) is introduced, it must have a negative coefficient. This means that the origin, where nothing is produced, cannot form the initial BFS since it is not feasible. At least 15 GJ of energy must be produced to prevent S_5 from taking a negative (ie infeasible) value.

The problem of finding a starting point may be overcome by introducing an artificial variable Q_1,

which represents an imaginary supply of pure, but very costly energy. Thus each unit of this activity is assumed to supply one GJ of energy and to cost, say, £1000, as shown in Table 16.6, Tableau I. Then 1000 times each element of the dietary constraint row must be subtracted from row 0, to give the initial BFS as shown in Tableau II. The problem can then be solved in the usual way. In this case the graphical method can be used as shown in Figure 16.4. The new constraint has caused a slight reduction in the optimal area of groundnuts and a small increase in the area of maize. Total gross margin is reduced to £485, and each GJ of energy has a shadow price of £12.

Of course, the actual cash earning at this point is zero, because all the produce is consumed. The objective of maximizing *total* gross margin is perhaps inappropriate given the subsistence constraint. If, instead, the objective were to maximize the *surplus* gross margin over subsistence needs, the optimal solution would be to grow two hectares of maize for subsistence, and to use the remaining July labour to produce ¾ hectare of groundnuts for sale (point N in

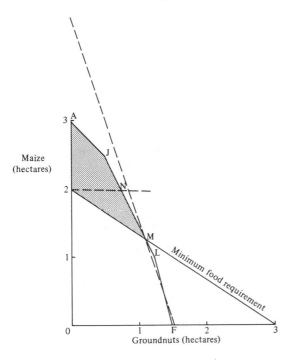

Fig. 16.4 Meeting minimum dietary constraints

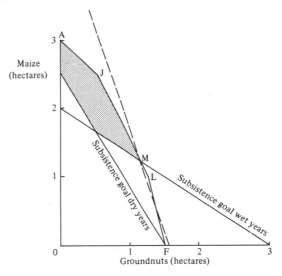

Fig. 16.5 Safety-first programming

Figure 16.4). For solution by the Simplex method, separate maize and groundnut selling activities would have to be incorporated with tie rows linking them to maize and groundnut production.

A useful extension of this approach, to allow for risk, is based on the inclusion of minimum survival constraints for each state of nature. To illustrate, let us assume that there are only two possible states of nature: maize yields 7.5 GJ of food energy per hectare in wet years but only 6 GJ in dry years. Groundnuts on the other hand yield 5 GJ in wet years but 10 in dry years. The constraint, equations are then

$$7.5X_1 + 5X_2 \geqslant 15 \text{ for wet years}$$

and

$$6X_1 + 10X_2 \geqslant 15 \text{ for dry years}$$

Slack and artificial variables can be added and the problem solved in the usual way. Also see the feasible area shown in Figure 16.5. It is a 'safety first' model in that the programme ensures that subsistence needs are met under all states of nature, before an optimum is sought (see Chapter 5). This method was used to analyse farm systems in South East Ghana. The resulting farm plan was closely similar to actual farming systems in the area, which was interpreted as evidence that farmers actually do make choices in this way. (See Low, 1974.)

Apart from planning for different states of nature, there may be several different goals or targets presented as constraints, for instance, minimal protein requirements for human nutrition, minimal feed requirements for livestock, or minimal cash requirements at certain periods of the year. Indeed, most 'basic needs' can be expressed in this form.

This leads on to the possibility of using LP to minimize the cost of meeting a given set of targets, rather than maximizing the return. A cost minimization problem can easily be converted into a maximization problem by changing the signs of the coefficients in the Z row. Minimizing C is just the same as maximizing minus C.

(iv) *Distinguishing between land types*

The example used above to illustrate LP includes separate constraints for labour at different periods of the year. Where there is division of labour between hired and family workers or between men and women, each category should be represented by a separate constraint for each peak work period.

Table 16.7. *Data for LP problem: example 2*

	Activities		
	Good land cropping (ha)	Poor land cropping (ha)	Constraint level
Yield of grain (quintals)	8	4	
Constraints			
Good land (hectares)	1	0	2
Poor land (hectares)	0	1	2
Total labour (man–days)	80	80	240

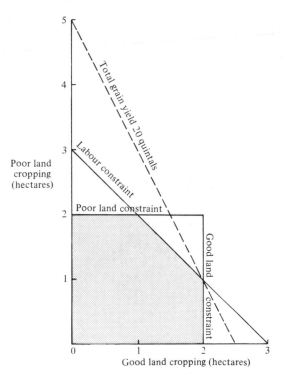

Fig. 16.6 Linear programming land use

In just the same way, where there are two cropping seasons, early season land should be treated as a separate constraint from late season land. Thus a 3 hectare farm has 3 hectares of early land and 3 hectares of late season land (see Clayton, 1961). Also, if there are different land types with different cropping alternatives for each, for example upland and river valley land, these should be treated as separate constraints.

Consideration of an alternative simple LP problem with just two types of land, good and poor, reinforces the arguments of Chapter 6, regarding intensity of land use. Example 2 is set out in Table 16.7. Note that we have assumed both land types require the same labour input per hectare but that the poor land produces half the yield of good land.

The problem is easily solved, as shown graphically in Figure 16.6. At the optimum, all the good land is in use but one hectare of the poor (marginal) land is unused. Twenty quintals of grain are produced. The shadow price of labour is 1/20 quintals or 5 kg, while the shadow price, or rents, of good and poor land are 4 quintals and zero respectively. Any increase in labour availability would increase the cultivated area of poor land, or extend the margin of cultivation.

(v) *Multiperiod programming*

LP is readily extended to allow for investment of either circulating or durable capital. If we treat time as being divided up into a series of discrete periods, eg cropping seasons, investment simply means that produce or purchases in one period are carried forward for use in a subsequent time period. Thus separate sets of activities and constraints are needed for each time period within the plan, while tie rows are used to represent investments, linking the various time periods together. The resultant 'block-diagonal' structure of the initial LP tableau is shown in Figure 16.7.

The only other consideration is the treatment of the objective function. Clearly produce which is invested in a given period cannot also be sold or consumed in that period. Hence consumption or sales activities in each period must compete with investment for the future. This is represented by the tie rows into which the current productive activities supply funds and from which the consumption activities and future production activities draw funds. The objective function then includes the values of current consumption activities *plus* the total value of what is produced in the terminal period.

We may illustrate with a simple, two period, example (3) with two productive activities, maize and groundnuts, and three constraints in each period,

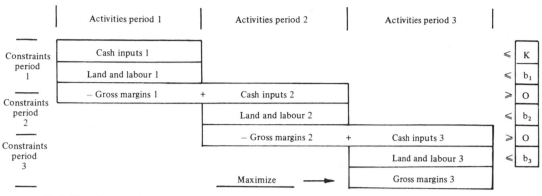

K = Initial cash reserves

b_i = Constraint levels in the ith period

Fig. 16.7 Multiperiod programming

land, labour and working capital. The data are set out in Table 16.8 and the initial tableau is shown in Table 16.9. Note that X_1 and X_2 now represent the areas of maize and groundnuts planned for year 1 while X_4 and X_5 represent the areas of maize and groundnuts planned for year 2. Activity X_3 simply allows for current consumption. Similar land and labour constraints apply in both years but whereas in year 1, cash needs are met from reserves of £200, in year 2 they must be met from year 1 income.

The optimal solution is obtained after four iterations as follows:

Total gross margin in second year	£608
year 1: maize	$2\frac{2}{3}$ hectares
groundnuts	$\frac{1}{3}$ hectare
unused labour	40 man–days
year 2: maize	$\frac{8}{15}$ hectare
groundnuts	$1\frac{11}{15}$ hectare
unused land	$\frac{11}{15}$ hectare

The corresponding shadow prices are

year 1:	Land (per ha)	£32.00
	Working Capital (per £)	£ 1.76
	Consumption (per £)	£ 1.00
year 2:	Labour (per man–day)	£ 0.67
	Working capital (per £)	£ 1.20

Thus we have a situation where, through shortage of working capital in the first year, the emphasis is on maize production. All the land is cultivated but some labour is left unused. By the second year more working capital is accumulated so the emphasis can shift to groundnuts, which are more profitable. However, they require more labour per hectare, so labour becomes the binding constraint and some land is left in disposal.

Note that the shadow price of £1 consumed is £1 as we might expect, but the shadow price of capital is £1.76 per £1 invested in year 1 and £1.20 per £1 invested in year 2. The rates of return are therefore 76 per cent and 20 per cent respectively. However, the rate of return of 76 per cent on year 1 capital is actually earned over two years. Given that the rate of

Table 16.8. *Data for LP problem: example 3*

	Activities		
	Maize	Groundnut	Constraint level
Annual gross margin	100	320	
Constraints			
Land (hectares)	1	1	3
Labour (man–days)	60	120	240
Capital (£)	50	200	200 (year 1)
			0 (year 2)

Table 16.9. *Initial tableau for 2 period LP*

		Z	X_1	X_2	X_3	S_1	S_2	S_3	X_4	X_5	S_4	S_5	S_6	Constant b
Row 0	Objective	1	0	0	-1	0	0	0	-100	-320	0	0	0	0
Row 1	Capital year 1	0	50	200	1	1	0	0	0	0	0	0	0	200
Row 2	land 1	0	1	1	0	0	1	0	0	0	0	0	0	3
Row 3	labour 1	0	60	120	0	0	0	1	0	0	0	0	0	240
Row 4	Capital year 2	0	-100	-320	0	0	0	0	50	200	1	0	0	0
Row 5	land 2	0	0	0	0	0	0	0	1	1	0	1	0	3
Row 6	labour 2	0	0	0	0	0	0	0	60	120	0	0	1	240

return in the second year is 20 per cent, the rate of return in the first year is obtained as follows:

$$(1 + r_1)(1 + r_2) = 1.76$$
$$1.2(1 + r_1) = 1.76$$
$$1 + r_1 = 1.76/1.2 = 1.467$$

Hence $r_1 = 46.7$ per cent.

This method of analysis can be applied to any multiperiod planning problem. The number of periods may be increased, using the same approach, while a 'period' may be less than or more than a year.

Furthermore, the 'working capital' transferred through the tie rows may be in cash or in kind. Thus maize storage might be represented by the transfer of grain from one seasonal period to the next.

Investment in durable capital, such as oxen and a plough, is represented by an investment activity. Such an activity is illustrated by the second column of Table 16.10 which is adapted from Table 16.9 by adding durable capital investment, borrowing and lending activities and omitting the disposal activities for ease of presentation. Of course, slack variables would have to be added to arrive at a solution. We assume that the durable capital investment can be treated as an integer variable, that it costs £150 initially, depreciates by 20 per cent to £120 over two years and saves 80 man–days of labour each year. The initial cost is set against cash funds in year 1, the terminal value appears in the objective function and the services it provides (in this case labour savings) appear in tie rows for each year of the plan. For the set of coefficients given, the investment is unlikely to

be attractive since the costs of £1.76 × 150 = £264 far exceed the benefits of £0.67 × 80 + £120 = £173.33. However, if credit is available the picture changes.

Borrowing and lending activities are also included in Table 16.10. A borrowing activity contributes to cash funds in one period but involves a cost of the loan plus interest (assumed to be 10 per cent here) in the following period. It may be necessary to include a constraint to reflect the limits on borrowing likely to apply in practice. A lending activity, on the other hand, has a cost in one period but yields a return of the principal plus interest in the next period. At 10 per cent interest, borrowing is clearly attractive and the optimal solution is to borrow £483.33 for the purchase of the durable asset and expansion of the groundnut area in year 1 to $2\frac{2}{3}$ hectares.

Thus we have a method of dynamic planning which can, in principle allow for all the complexities of intertemporal choice. Alternative investment opportunities with different productive lives and different starting dates may be considered, together with a range of alternative borrowing and lending activities. The main limitation is that of scale. Since a matrix of coefficients is included for each period of the plan, the whole multi-period problem may become very large indeed. Clearly, the choice of time horizon, affects the size of the problem. but some reduction may be achieved by including activities for, say, every fifth year, rather than for all years. Skill and judgement are involved in deciding the amount of detail and complexity needed to provide an adequate representation of the real world problem.

Table 16.10. *Initial tableau example 4: Durable capital, borrowing and lending*

		Activities								
	Z	Durable* capital investment	Maize year 1	G'nuts year 1	Consume year 1	Borrow year 1	Maize year 2	G'nuts year 2	Lend year 2	Constant b
Objective	1	−120	0	0	−1	0	−100	−320	−1.10	0
Cash year 1	0	150	50	200	1	−1	0	0	0	200
Land year 1	0	0	1	1	0	0	0	0	0	3
Labour year 1	0	−80	60	120	0	0	0	0	0	240
Cash year 2	0	0	−100	−320	0	1.10	50	200	1	0
Land year 2	0	0	0	0	0	0	1	1	0	3
Labour year 2	0	−80	0	0	0	0	60	120	0	240

* Integer variable; 20 per cent depreciation over two years; saves 80 man–days per year.

(vi) *Choice of objectives*

It may have been noted in the previous section that first-year consumption did not appear in the optimum solution; the entire gross margin from the first year was allocated to productive investment for the second year. This is the inevitable result of the way the objective function was formulated, giving equal weightings to current and future consumption. Thus lending even at very low interest rates yields more than consumption. The total consumption is then maximized by maximizing growth. Greater realism might be achieved by imposing minimum subsistence consumption constraints for each year of the plan, to allow for basic food needs. However, growth would still be maximized subject to these constraints.

If, as seems more likely, farmers are willing to forego some growth in exchange for somewhat higher current consumption we must discount future returns in the objective function. The appropriate discount rate to use is the estimated personal discount rate of the farmer. This choice is important because it establishes the threshold rate of return for the inclusion of productive activities in the plan. In our example, given a personal discount rate of 25 per cent, the value for consumption in the first year becomes $-1/1.25$, while all the returns in the second year must be divided by 1.25^2. Thus current consumption becomes more attractive relative to investment for the future.

The objective is now to maximize net present value of annual consumption flows plus the terminal assets. Investment appraisal is effectively incorporated within an LP framework. This approach was used to plan tree-crop investment in East and West Cameroon by Abalu (1975).

Further refinements are possible, in specifying the objective function. For instance the value or utility per unit of consumption may decline with increasing consumption in a particular year. This is easily represented by including more than one consumption activity in each year. For illustration, each £1 of consumption of the first £100 might be valued at 1 unit of utility. Beyond this each £1 is valued at 0.5 units. Apart from the three activities, in each year, two additional constraints would be needed, limiting each of the first two consumption activities to a maximum of £100.

A broadly similar approach has been proposed to deal with risk rather than multiperiod planning. In this method, known as discrete stochastic programming each 'block' of the tableau relates to a particular state of nature and the values in the objective function are 'expected utilities' obtained as the product of the probability of a particular state of nature and the utility of the consumption on a diminishing scale as described above.

However, as discussed earlier in Chapters 4 and 5, rather than attempting to combine multiple objectives into a single utility function it might be more practical simply to rank goals in order of priority. High priority goals may then be treated as targets or constraints. A more sophisticated approach which allows deviations from the target levels of achievement of various objectives is known as 'goal programming' (see Romero & Rehman, 1984; 1985). One application of goal programming to farm planning in Senegal gives results which only differ slightly from those of ordinary linear programming (Barnett, Blake & McCarl, 1982).

Practical application of LP

It should be clear from this discussion that LP is a very flexible tool. Many different on- and off-farm activities can be included, physical and social constraints can be imposed, and it can be adapted to deal with multiple objectives, risk and planning over time. However, its most valuable feature is that it is a whole farm planning tool, particularly suited to handling the interactions of a complex system. In contrast, partial-budgeting is concerned only with one activity or sub-system.

The main disadvantage of LP is its complexity and the need to use a computer for all but the simplest problems. As already remarked, skill and judgement are needed to assess how complex the model should be in order to provide useful insights. However, even very simple versions, such as those described in this chapter can help our understanding of the general problem of optimization subject to constraints. The increasing availability of computers and LP packages for microcomputers allows wider use of this tool by researchers and farm advisers.

Nevertheless, the use of LP is too costly for individual farm planning, except perhaps for very large-scale units such as State Farms. The main applications are based on analysis of a typical, average or modal farm which is supposed to represent the whole population. Thus it is used either to predict how the majority of these farmers will respond to technical

and economic changes, or to prescribe how their farming systems might be improved. Clearly, this raises problems of defining the target population and identifying a representative case study. Furthermore, the value of the results depends upon the accuracy and reliability of the data. But these problems arise with any objective approach to farming systems research, as discussed in Chapter 10.

Another problem that arises with any kind of model building, or indeed theorizing about farming systems, is that of validation. We start with assumptions or hypotheses about farmer objectives and subjective constraints and base our predictions and prescriptions on these. But it is very difficult to test whether our initial assumptions are valid. Validation procedures are generally based on comparing the model predictions with what farmers are actually doing. If the predictions differ from reality we conclude the initial assumptions were wrong and need modifying, but it could alternatively be concluded that the assumptions were right and farm management practices should be changed. If the predictions are accurate, on the other hand, we cannot be sure that the assumptions were correct. However, if the model is tested and validated over a range of circumstances our confidence in its use is strengthened. Farmers may of course be asked what their objectives and constraints are, but their replies may be too vague and ill-formed for inclusion in quantitative models.

The criticism has been made that LP, even multi-period LP, is suited only to planning at a specific point in time, whereas farmers are sequential decision makers who continuously adapt to changing circumstances (see Chapter 5). However, this is hardly a valid criticism. After all, sequential decision making simply involves a sequence of plans made at specific points in time. In view of this, farmers maintain flexibility, for instance by leaving some land uncleared until the rains are well established. Suitable constraints could be included in an LP model to represent this practice. However, the main point is that at the start of the season the farmer must make a comprehensive plan for the whole season, possibly including some sequential clearing and planting. Later, perhaps only a few days later, he may revise the plan in the light of the weather and other external variables, but that does not alter the fact that he started with a comprehensive plan, which can be modelled using LP. Used intelligently this planning tool can yield valuable insights into the structure and possible improvements of farming systems.

Further reading

Abalu, G. O. I. (1975). Optimal investment decisions in perennial crop production: a dynamic linear programming approach, *Journal of Agricultural Economics*, **26**(3) 383

Ahmed, I. & Kinsey, B. H. (eds) (1984). *Farm Equipment Innovations in Eastern and Central Southern Africa*, Aldershot, Hants. Gower for ILO

Barnett, D., Blake, B. & McCarl, B. A. (1982) Goal programming via multidimensional scaling applied to Senegalese subsistence farms, *American Journal of Agricultural Economics* **64**(4) 720–7

Clayton, E. S. (1961). Technical and economic optima in peasant agriculture, *Journal of Agricultural Economics* **14**(3) 337

Farrington, J. (1976). A note on planned versus actual farmer performance under uncertainty in underdeveloped agriculture, *Journal of Agricultural Economics*, **27**(2) 257

Hardaker, J. B. (1979). A review of some farm management research methods for small farm development in LDCs, *Journal of Agricultural Economics*, **30**(3) 315

Heyer, J. (1972). An analysis of peasant farm production under conditions of uncertainty, *Journal of Agricultural Economics*, **23**(2) 135

Knipscheer, H. C., Menz, K. M. & Verinumbe, I. (1983). The evaluation of preliminary FS technologies: zero tillage systems in W. Africa, *Agricultural Systems*, **11**(2) 95–103

Low, A. R. C. (1974). Decision taking under uncertainty: a linear programming model of peasant farmer behaviour, *Journal of Agricultural Economics*, **25**(3) 311

Ogunfowora, O. (1972). Conceptualizing increased resource demand and product supply inducing policies in peasant agriculture, *Nigerian Journal of Economic and Social Studies*, **14**(2) 191

Ogunfowora, O. & Norman, D. W. (1973). Farm-firm normative fertilizer demand response in the North Central State of Nigeria. *Journal of Agricultural Economics*, **24**(2) 301

Ogunfowora, O. & Norman, D. W. (1974). *An optimization model for evaluating the stability of sole cropping and mixed cropping systems under changing resource and technology levels*, Zaria, Nigeria, Ahmadu Bello Univ, IAR. Research Bulletin No. 217

Olayide, S. O. & Olowude, S. O. (1972). Optimum combination of farm enterprises in W. Nigeria: a LP analysis, *Nigerian Journal of Economic and Social Studies*, **14**(1) 63

Romero, C. & Rehman, T. (1984). Goal programming and multiple criteria decision making in farm planning, *Journal of Agricultural Economics*, **35**(2) 177

Romero, C. & Rehman, T. (1985). Goal programming and multiple criteria decision making in farm planning: some extensions. *Journal of Agricultural Economics*, **36**(2) 171

Vail, D. J. (1973). Induced farm innovation and derived scientific research strategy, *East African Journal of Rural Development*, **6**(1)

Appendix

Table I. Present value (P) of £1 due in n years. $P = \dfrac{1}{(1+i)^n}$

							Rate of interest (i) per cent								
n	2.00	4.00	6.00	8.00	10.00	12.00	14.00	16.00	18.00	20.00	22.00	24.00	26.00	28.00	30.00
1	0.9804	0.9615	0.9434	0.9259	0.9091	0.8929	0.8772	0.8621	0.8475	0.8333	0.8197	0.8065	0.7937	0.7813	0.7692
2	0.9612	0.9246	0.8900	0.8573	0.8264	0.7972	0.7695	0.7432	0.7182	0.6944	0.6719	0.6504	0.6299	0.6104	0.5917
3	0.9423	0.8890	0.8396	0.7938	0.7513	0.7118	0.6750	0.6407	0.6086	0.5787	0.5507	0.5245	0.4999	0.4768	0.4552
4	0.9238	0.8548	0.7921	0.7350	0.6830	0.6355	0.5921	0.5523	0.5158	0.4823	0.4514	0.4230	0.3968	0.3725	0.3501
5	0.9057	0.8219	0.7473	0.6806	0.6209	0.5674	0.5194	0.4761	0.4371	0.4019	0.3700	0.3411	0.3149	0.2910	0.2693
6	0.8880	0.7903	0.7050	0.6302	0.5645	0.5066	0.4556	0.4104	0.3704	0.3349	0.3033	0.2751	0.2499	0.2274	0.2072
7	0.8706	0.7599	0.6651	0.5835	0.5132	0.4523	0.3996	0.3538	0.3139	0.2791	0.2486	0.2218	0.1983	0.1776	0.1594
8	0.8535	0.7307	0.6274	0.5403	0.4665	0.4039	0.3506	0.3050	0.2660	0.2326	0.2038	0.1789	0.1574	0.1388	0.1226
9	0.8368	0.7026	0.5919	0.5002	0.4241	0.3606	0.3075	0.2630	0.2255	0.1938	0.1670	0.1443	0.1249	0.1084	0.0943
10	0.8203	0.6756	0.5584	0.4632	0.3855	0.3220	0.2697	0.2267	0.1911	0.1615	0.1369	0.1164	0.0992	0.0847	0.0725
11	0.8043	0.6496	0.5268	0.4289	0.3505	0.2875	0.2366	0.1954	0.1619	0.1346	0.1122	0.0938	0.0787	0.0662	0.0558
12	0.7885	0.6246	0.4970	0.3971	0.3186	0.2567	0.2076	0.1685	0.1372	0.1122	0.0920	0.0757	0.0625	0.0517	0.0429
13	0.7730	0.6006	0.4688	0.3677	0.2897	0.2292	0.1821	0.1452	0.1163	0.0935	0.0754	0.0610	0.0496	0.0404	0.0330
14	0.7579	0.5775	0.4423	0.3405	0.2633	0.2046	0.1597	0.1252	0.0985	0.0779	0.0618	0.0492	0.0393	0.0316	0.0254
15	0.7430	0.5553	0.4173	0.3152	0.2394	0.1827	0.1401	0.1079	0.0835	0.0649	0.0507	0.0397	0.0312	0.0247	0.0195
16	0.7284	0.5339	0.3936	0.2919	0.2176	0.1631	0.1229	0.0930	0.0708	0.0541	0.0415	0.0320	0.0248	0.0193	0.0150
17	0.7142	0.5134	0.3714	0.2703	0.1978	0.1456	0.1078	0.0802	0.0600	0.0451	0.0340	0.0258	0.0197	0.0150	0.0116
18	0.7002	0.4936	0.3503	0.2502	0.1799	0.1300	0.0946	0.0691	0.0508	0.0376	0.0279	0.0208	0.0156	0.0118	0.0089
19	0.6864	0.4746	0.3305	0.2317	0.1635	0.1161	0.0829	0.0596	0.0431	0.0313	0.0229	0.0168	0.0124	0.0092	0.0068
20	0.6730	0.4564	0.3118	0.2145	0.1486	0.1037	0.0728	0.0514	0.0365	0.0261	0.0187	0.0135	0.0098	0.0072	0.0053
21	0.6598	0.4388	0.2942	0.1987	0.1351	0.0926	0.0638	0.0443	0.0309	0.0217	0.0154	0.0109	0.0078	0.0056	0.0040
22	0.6468	0.4220	0.2775	0.1839	0.1228	0.0826	0.0560	0.0382	0.0262	0.0181	0.0126	0.0088	0.0062	0.0044	0.0031
23	0.6342	0.4057	0.2618	0.1703	0.1117	0.0738	0.0491	0.0329	0.0222	0.0151	0.0103	0.0071	0.0049	0.0034	0.0024
24	0.6217	0.3901	0.2470	0.1577	0.1015	0.0659	0.0431	0.0284	0.0188	0.0126	0.0085	0.0057	0.0039	0.0027	0.0018
25	0.6095	0.3751	0.2330	0.1460	0.0923	0.0588	0.0378	0.0245	0.0160	0.0105	0.0069	0.0046	0.0031	0.0021	0.0014
26	0.5976	0.3607	0.2198	0.1352	0.0839	0.0525	0.0331	0.0211	0.0135	0.0087	0.0057	0.0037	0.0025	0.0016	0.0011
27	0.5859	0.3468	0.2074	0.1252	0.0763	0.0469	0.0291	0.0182	0.0115	0.0073	0.0047	0.0030	0.0019	0.0013	0.0008
28	0.5744	0.3335	0.1956	0.1159	0.0693	0.0419	0.0255	0.0157	0.0097	0.0061	0.0038	0.0024	0.0015	0.0010	0.0006
29	0.5631	0.3207	0.1846	0.1073	0.0630	0.0374	0.0224	0.0135	0.0082	0.0051	0.0031	0.0020	0.0012	0.0008	0.0005
30	0.5521	0.3083	0.1741	0.0994	0.0573	0.0334	0.0196	0.0116	0.0070	0.0042	0.0026	0.0016	0.0010	0.0006	0.0004

Table II. Present value (P_{an}) of an annuity of £1 payable for n future years. $P_{an} = \dfrac{[(1+i)^n - 1]}{i(1+i)^n}$

n	\multicolumn{15}{c}{Rate of interest (i) per cent}
	2.00	4.00	6.00	8.00	10.00	12.00	14.00	16.00	18.00	20.00	22.00	24.00	26.00	28.00	30.00
1	0.9804	0.9615	0.9434	0.9259	0.9091	0.8929	0.8772	0.8621	0.8475	0.8333	0.8197	0.8065	0.7937	0.7813	0.7692
2	1.9416	1.8861	1.8334	1.7833	1.7355	1.6901	1.6467	1.6052	1.5656	1.5278	1.4915	1.4568	1.4235	1.3916	1.3609
3	2.8839	2.7751	2.6730	2.5771	2.4869	2.4018	2.3216	2.2459	2.1743	2.1065	2.0422	1.9813	1.9234	1.8684	1.8161
4	3.8077	3.6299	3.4651	3.3121	3.1699	3.0373	2.9137	2.7982	2.6901	2.5887	2.4936	2.4043	2.3202	2.2410	2.1662
5	4.7135	4.4518	4.2124	3.9927	3.7908	3.6048	3.4331	3.2743	3.1272	2.9906	2.8636	2.7454	2.6351	2.5320	2.4356
6	5.6014	5.2421	4.9173	4.6229	4.3553	4.1114	3.8887	3.6847	3.4976	3.3255	3.1669	3.0205	2.8850	2.7594	2.6427
7	6.4720	6.0021	5.5824	5.2064	4.8684	4.5638	4.2883	4.0386	3.8115	3.6046	3.4155	3.2423	3.0833	2.9370	2.8021
8	7.3255	6.7327	6.2098	5.7466	5.3349	4.9676	4.6389	4.3436	4.0776	3.8372	3.6193	3.4212	3.2407	3.0758	2.9247
9	8.1622	7.4353	6.8017	6.2469	5.7590	5.3282	4.9464	4.6065	4.3030	4.0310	3.7863	3.5655	3.3657	3.1842	3.0190
10	8.9826	8.1109	7.3601	6.7101	6.1446	5.6502	5.2161	4.8332	4.4941	4.1925	3.9232	3.6819	3.4648	3.2689	3.0915
11	9.7868	8.7605	7.8869	7.1390	6.4951	5.9377	5.4527	5.0286	4.6560	4.3271	4.0354	3.7757	3.5435	3.3351	3.1473
12	10.5753	9.3851	8.3838	7.5361	6.8137	6.1944	5.6603	5.1971	4.7932	4.4392	4.1274	3.8514	3.6059	3.3868	3.1903
13	11.3484	9.9856	8.8527	7.9038	7.1034	6.4235	5.8424	5.3423	4.9095	4.5327	4.2028	3.9124	3.6555	3.4272	4.2233
14	12.1062	10.5631	9.2950	8.2442	7.3667	6.6282	6.0021	5.4675	5.0081	4.6106	4.2646	3.9616	3.6949	3.4587	3.2487
15	12.8493	11.1184	9.7122	8.5595	7.6061	6.8109	6.1422	5.5755	5.0916	4.6755	4.3152	4.0013	3.7261	3.4834	3.2682
16	13.5777	11.6523	10.1059	8.8514	7.8237	6.9740	6.2651	5.6685	5.1624	4.7296	4.3567	4.0333	3.7509	3.5026	3.2832
17	14.2919	12.1657	10.4773	9.1216	8.0216	7.1196	6.3729	5.7487	5.2223	4.7746	4.3908	4.0591	3.7705	3.5177	3.2948
18	14.9920	12.6593	10.8276	9.3719	8.2014	7.2497	6.4674	5.8178	5.2732	4.8122	4.4187	4.0799	3.7861	3.5294	3.3037
19	15.6785	13.1339	11.1581	9.6036	8.3649	7.3658	6.5504	5.8775	5.3162	4.8435	4.4415	4.0967	3.7985	3.5386	3.3105
20	16.3514	13.5903	11.4699	9.8181	8.5136	7.4694	6.6231	5.9288	5.3527	4.8696	4.4603	4.1103	3.8083	3.5458	3.3158
21	17.0112	14.0292	11.7641	10.0168	8.6487	7.5620	6.6870	5.9731	5.3837	4.8913	4.4756	4.1212	3.8161	3.5514	3.3198
22	17.6580	14.4511	12.0416	10.2007	8.7715	7.6446	6.7429	6.0113	5.4099	4.9094	4.4882	4.1300	3.8223	3.5558	3.3230
23	18.2922	14.8568	12.3034	10.3711	8.8832	7.7184	6.7921	6.0442	5.4321	4.9245	4.4985	4.1371	3.8273	3.5592	3.3254
24	18.9139	15.2470	12.5504	10.5288	8.9847	7.7843	6.8351	6.0726	5.4509	4.9371	4.5070	4.1428	3.8312	3.5619	3.3272
25	19.5235	15.6221	12.7834	10.6748	9.0770	7.8431	6.8729	6.0971	5.4669	4.9476	4.5139	4.1474	3.8342	3.5640	3.3286
26	20.1210	15.9828	13.0032	10.8100	9.1609	7.8957	6.9061	6.1182	5.4804	4.9563	4.5196	4.1511	3.8367	3.5656	3.3297
27	20.7069	16.3296	13.2105	10.9352	9.2372	7.9426	6.9352	6.1364	5.4919	4.9636	4.5243	4.1542	3.8387	3.5669	3.3305
28	21.2813	16.6631	13.4062	11.0511	9.3066	7.9844	6.9607	6.1520	5.5016	4.9697	4.5281	4.1566	3.8402	3.5679	3.3312
29	21.8444	16.9837	13.5907	11.1584	9.3696	8.0218	6.9830	6.1656	5.5098	4.9747	4.5312	4.1585	3.8414	3.5687	3.3317
30	22.3965	17.2920	13.7648	11.2578	9.4269	8.0552	7.0027	6.1772	5.5168	4.9789	4.5338	4.1601	3.8424	3.5693	3.3321

Index

objectives, management, 5, 13–14, 37, 91–2, 101, 106, 145–6, 177–8
observation, direct, 107, 114
off-farm work activities, 4, 47, 74–5, 99
offsets for land survey, 115
offtake, harvested, 57
on-farm testing, 102, 105, 108, 122
open access, 61–2
opening valuation, 118
operating costs, 30
opportunity cost, 34, 38, 43, 59–60, 69–70, 74, 78, 81–4, 87, 105, 126, 137, 156, 165, 170
optimal solution (*see also* economic optimum), 149, 164, 167
organic fertilizer (*see also* manure), 138
  matter, 7–8
outerbound production function, 90
outgrower schemes, 31
output, 19, 23
  effect, 21
  measurement, 124–5
overcropping, 55
overexploitation, 57
overgrazing, 55
overheads, 125
oxen (*see also* animal draught, bullocks), 75, 138, 176

parametric programming, 170–1
partial budgets, 145–6, 150–3, 170
participatory research, 103
part-time farming, 75–6
parturition interval and rate, 127
pastoralism, 4, 7–8, 12, 31–2, 55, 58, 63, 150
pastoral systems research, 100
pay-off matrix, 41
peak, labour, 68, 75, 152
  working capital, 82
perennial, permanent crops, 5, 10, 31, 105, 127
permanent cultivation, 58, 61
perpetuity, 155
personal discount rate, 81, 177
pests and diseases, 8
pesticides, 51, 65
piece work, 73
pilot survey, 113
pivot, linear programming, 168
plane table for land survey, 115
planning, farm 144–79
  horizon, 154–5
plantations (*see also* estates), 31, 94
planting date, 7
plant nutrients, 7, 34–5
pledge of land, 63–4
pluri-activity, agricultural (*see also* part-time farming), 4
population growth, 11, 28, 58, 62–3, 85
  support capacity, 8–9
precision of estimates, 108, 110–11

prediction and prescription, 131, 145
present value, 83, 118, 155, 180–1
price, 10–13, 30
  change, effects of, 5, 11, 18, 21, 38–9
  incentives, 11, 13, 99
  policy, 5, 13
  response, 5, 11
price stability, 51
  uncertainty and variation, 41
principle of increasing risk, 86
priority of objectives (*see also* lexicographic ordering), 37
privatization of natural resources, 63
probability, 42–6, 49, 108
  conditional, 47
  distribution, 121
processing, 31, 116
product combinations, 33
  price (*see also* marginal revenue), 26
  transformation, rate of, 33
production functions, 15, 90, 129–41, 149
  possibility boundary, curve or frontier, 33, 35, 38, 80, 164
productive resources, 5
productivity, 13, 16–17
profit maximization, 14, 17–18, 26, 146
programme planning, 153
project appraisal, 154
purchases and sales, 106

quadratic function, 45, 132, 136
  programming, 45
questionnaires, 111–12, 120
quick and dirty methods, 101, 103–4

rainfall, 6–7, 55
ranches, 31
random sampling, 109, 122
rangelands, 6–7, 56, 61
ranking objectives, 37, 44
rapid rural appraisal (*see also* quick and dirty methods), 101, 107
rate of interest, 86–7
  of product transformation, 33
  of return, 80
  of technical substitution, 19
  of working, 67
rationality, farmer, 14, 42
real activity, 168
recall, 106–7, 114
receipts (*see also* sales), 126
reciprocity, 9, 67
recommendation domain, 100, 104, 145
reconnaissance or pilot survey, 109, 113
records, 106, 114
  input, 145
  sales, 117–18
regeneration of fertility, 8, 57

SOCIAL SCIENCE LIBRARY

Oxford University Library Services

Manor Road

Oxford OX1 3UQ

Tel: (2)71093 (enquiries and renewals)

http://www.ssl.ox.ac.uk

WITHDRAWN

This is a NORMAL LOAN item.

We will email you a reminder before this item is due.

Please see http://www.ssl.ox.ac.uk/lending.html
for details on:

- loan policies; these are also displayed on the notice boards and in our library guide.

- how to check when your books are due back.

- how to renew your books, including information on the maximum number of renewals. Items may be renewed if not reserved by another reader. Items must be renewed before the library closes on the due date.

- level of fines; fines are charged on overdue books.

Please note that this item may be recalled during Term.